BY JIM ROBBINS

*The Wonder of Birds*
*The Man Who Planted Trees*

# The Wonder of Birds

# The Wonder of Birds

What They Tell Us About Ourselves,
the World, and a Better Future

## JIM ROBBINS

SPIEGEL & GRAU

NEW YORK

Published in the United States by Spiegel & Grau,
an imprint of Random House, a division of
Penguin Random House LLC, New York.

Spiegel & Grau and Design is a registered trademark of
Penguin Random House LLC.

LIBRARY OF CONGRESS CATALOGING-IN-PUBLICATION DATA
Names: Robbins, Jim, author.
Title: The wonder of birds / by Jim Robbins.
Description: First edition. | New York : Spiegel & Grau, 2017. |
Includes bibliographical references and index.
Identifiers: LCCN 2016049366 | ISBN 9780812993530 |
ISBN 9780679645672 (ebook)
Subjects: LCSH: Birds.
Classification: LCC QL676 .R63 2017 | DDC 598—dc23
LC record available at lccn.loc.gov/2016049366

Printed in the United States of America on acid-free paper

randomhousebooks.com
spiegelandgrau.com

2 4 6 8 9 7 5 3 1

FIRST EDITION

Book design by Dana Leigh Blanchette

For Chere

For Betty
With much love and gratitude

For Jim,
who has been called home

For Matthew and Annika,
who keep me laughing

And for the birds,
who never let me down

Look deep into Nature and then
you will understand everything better.

—ATTRIBUTED TO ALBERT EINSTEIN

# CONTENTS

# PART II

## The Gifts of Birds

# PART III

## Discovering Ourselves Through Birds

# PART IV

## Birds and the Hope for a Better Future

# PREFACE

No wild animal lives so freely and in such variety and numbers among humans as do birds. For that reason alone, our relationship to them is unlike our connection to any other wild creature. But there are other reasons, too. The intellect of birds is arguably the closest in the animal world to our own. Birds charm us with their ethereal songs, which are profoundly different from the sound of any other animal; in fact, some of the natural world's most beautiful sounds emerge from the tiniest of birds. They are found virtually everywhere, from the Arctic and Antarctic to the tropics and deserts to the concrete labyrinths at the heart of the world's cities and the green patches of grass in front of our homes, and they are nature's exclamation point, adding an unequaled burst of vibrancy to our lives. Birds came to the earth, an Australian legend has it, when a rainbow shattered and its shards of color turned into birds as they fell: the glowing, jewel-like reds, greens, and blues of the humming-birds; the bold red, white, and black of the woodpeckers; the deep blue of bluebirds and indigo buntings; the slash of red on the shoulders of red-winged blackbirds and the full suit of red worn by cardinals.

I am in awe of birds. I knew something about them going into this project, yet after more than two years of reading scientific studies, talking with scientists and laymen, and visiting winemakers, zookeepers, bird-watchers, falconers, artists, costume designers, Native Americans, and animal activists, I discovered that these feathered creatures play an almost unfathomably wide range of roles in the human enterprise.

My interest lies in how we have interpreted birds, and how we might enhance our interpretation to help us understand more about ourselves and the world around us. For centuries, for example, birds have been used to help us understand flight and to create new designs for better and more efficient aircraft. Running and climbing baby partridges offer a glimpse into how the first flying dinosaurs left the ground, and the molecular tracings through the brain of a singing baby bird are a proxy for understanding the biology of the first babbles from a human infant. Want to know how Machiavellian power relationships among humans evolved? Look to the deceptions and alliances that take place within a flock of cunning ravens. Why are human stepfamilies more at risk for incest and other dysfunction than biological families? Families of white-fronted bee-eaters in the desert cliffs of Kenya, who are no strangers to illicit desire, can help answer this question. Murmurations, those mysterious and synchronous flying clouds of birds that mesmerize us as they swirl through a late afternoon sky in autumn, might hold the secret to an undiscovered fundamental force in physics.

One of the most important things birds do is remind us of our deep and abiding emotional connection to nature. In Washington, D.C., I witnessed Harris's hawks and bald eagles turn drug dealers and high school dropouts into passionate falconers. What is going on in our hearts and brains when we observe these creatures? What moves people to spend hundreds of dollars a year feeding birds in their backyard, or thousands to travel the world to watch them? Throughout history, birds have been strongly allied with mystical properties. Might birds, then, also have things to tell us that science has yet to consider? Birds push the boundaries of science, for example, by raising the big question of whether these highly complex beings have minds of their own. The idea of a sentient world beyond the human realm is not as far out as it has been thought to be. Panpsychism (literally, "mind everywhere"), the idea that the world and everything in it is conscious, is one of philosophy's oldest doctrines. It held sway until the twentieth century, when "logical" thinking took over and asserted the notion of human exceptionalism. But some scientists are rethinking this notion.

How can we divine an understanding of such a conscious world? Largely through birds. Ethno-ornithologists who study the relationship between the world's indigenous people and birds, and also the few scientists whose work at times questions the fundamental assumptions that we believe control the workings of our universe, tell us tales of alternative worlds and radically different modes of perception in which people are woven into the fabric of nature and birds are not just objects but fellow travelers, and sometimes even family.

Our society is unparalleled in understanding and appreciating material science, accumulating hard drives, clouds, and folders full of knowledge. But knowledge breaks the world into pieces, while wisdom makes it whole. Bird flight can be about air currents and velocity and wing loading, but it can also be about something unquantifiable, something transcendent and miraculous. We are sorely deficient in wisdom, lacking a view of how all of these pieces of knowledge connect to make the world a better place.

"What we see is not nature," the physicist Werner Heisenberg cautioned, "but nature exposed to our method of questioning." Birds, as understood through the field of ethno-ornithology, tell us that the Western scientific model is only one narrow porthole through which to envision the world; there are many more radically different perspectives, each with its own validity. As Heisenberg suggests, we think we know how the world works, but that belief is a dangerous illusion. It's time for us to stop laboring and debating solely within the framework of our own culture, our own model. It's time to be more inclusive and more creative, to peer beneath the surface, to vastly broaden our scope, to consider other ways of seeing and being in the world, and to reframe our perspective of nature, which will ultimately make us a far more resilient species.

"In the end we will only conserve what we love, we will love what we understand, and we will understand what we are taught," wrote Baba Dioum, a Senegalese forester. This book is my humble attempt to write about how a wide range of people interpret birds and to offer a few interpretations of my own, to teach something about this marvelous planet we call home and the fellow travelers with whom we share it, creatures who are able to fly halfway across the globe nonstop, dive ten times deeper into the ocean than a human, or fly backward and upside down and do many other things we cannot begin to comprehend.

My goal with this book is to help change the way we perceive birds, to move them from the background of our lives to the foreground, from the quotidian to the miraculous. In it I hope to share my own soul-stirring wonder, and I hope that that will be infectious. A shared wonder about the miraculous nature of birds may be the best strategy to reshape our relationship with these creatures, with the earth, and with ourselves.

PART I

What Birds Tell Us About
the Natural World

# CHAPTER 1

# Birds: The Dinosaurs That Made It

What good is half a wing?

—ST. GEORGE JACKSON MIVART

Where did the first flying bird come from? Did it spring, fully formed, with perfect wings, from the mind of God? Or was the first act of flying carried out by a small dinosaur with feathers who leaped out of a tree, glided gently through the air, and landed on the ground like a child's balsa airplane? Did a galloping feathered dinosaur chase so fast after a buzzing insect, leaping to gobble it, that it found itself airborne? How flight first happened is a mystery, but in the birds that surround us today, which are the only surviving dinosaur lineage, some have found a look back at how dinosaurs might have gone airborne and what these creatures from long ago were like.

The governing theories about how the first animal evolved the ability to fly—first proposed in the nineteenth century and still operable today—are divided into two main camps, the *arboreal* and the *cursorial*. Derived from the Latin word for "tree," the arboreal theory holds that around 125 million years ago, small reptilelike creatures with four limbs were covered with something like feathers. The featherlike covering was used not for flying but as a cloak to keep the creatures warm, or as a way to look sexy and attract partners, or as camouflage, or all three. Perhaps these creatures leaped from tree to tree in a dense rainforest, the way a flying squirrel travels—not really flying, but gliding.

Then one day, arborealists imagine, with its forelimbs stretched in front of it and its feather coverings spread out to the side, the first flying animal glided from a tree to the ground, and as it went on it added flapping to increase thrust. Perhaps the critter had a random genetic mutation that gave it larger forelimbs than others, which helped propel the animal forward. There are some sticky problems, however, that some argue shoot down the trees-to-ground theory, one of which is that there are no gliding animals today that flap for thrust.

The cursorial, or ground-up, theory of the origin of flight refers to the animal's ability to run. In this scenario, the first fliers were track stars with a yearning to take to the skies and soar. After zooming along the ground and making a series of leaps, to chase a dragonfly perhaps, or cross a creek, they somehow found themselves soaring with feathery forelimbs that had, perhaps through random mutations, grown large and light enough to keep them aloft. Left unexplained is where a heavy dinosaur would get the energy to run three times faster than modern birds in order to break the bonds of gravity and flap its way through the air, all with a developing wing. Perhaps, some have thought, they were half wings on a bipedal, or two-legged, creature, but why would an animal have a half wing if only a full wing would allow it to fly?

It's a good question, and one that's often asked when it comes to the evolution of flight. "What good is half a wing?" was first asked in 1871 by St. George Jackson Mivart, an English biologist. Mivart was at first devoted to Darwin's theory of natural selection—the idea that as creatures evolved, only those who were most fit survived. He turned against his mentor, though, and later became one of the theory's most vehement critics—largely over the bird wing. There is no reason on God's green earth for an animal to have half wings, because they are useless, he claimed. Ergo, the theory of evolution doesn't make sense; God must have created birds fully formed. To this day, creationists hold fast to the argument that half a bird wing refutes evolution.

This is where the first-flying-creature debate has stood for quite a while—two main schools of thought arguing about their respective ideas, with a separate school believing that a bird's wing is a result of the act of divine creation, rather than meticulous and persistent shaping by eons of evolution.

A new perspective that combined aspects of both evolutionary theories arose when Ken Dial weighed in in the early 2000s. While his distinguished career has been about studying bird flight mechanics, he found that his approach could also be uniquely applied to the evolution of flight. "Study the dinosaurs that made it—the birds," he says. Understanding more about the evolution of flight by studying living animals provides a new perspective on the ecology and biology of birds and dinosaurs, information that can't be gotten elsewhere. "These are things you would never get from studying fossils," he says.

With his shaved head, goatee, and glistening aviator sunglasses, the guitar-playing, jet-piloting Dial is perfect for the role of a renegade. He's a bird nut, as many people who investigate birds are, energetic and excited when talking about his research. The fact that he's an interloper on the subject, wading as a biologist into a field occupied largely by paleontologists, doesn't bother him. The origin-of-flight theorists base their respective arguments on the study of fossilized dinosaur bones. This necessarily involves a lot of conjecture because the beasts are so long gone, and their bodies are very unbirdlike at this point because they are fossils frozen in stone.

Dial's work is based on videos of hundreds of live birds performing in his lab, doing things that no one knew birds did, as well as his study of their muscles, limbs, and other mechanics. I watched several of these films with Dial in his office at the University of Montana Flight Laboratory in Missoula, and I asked him how he thinks bird flight first took off. His is a fascinating idea, based in part, he told me, on something called "recapitulation theory," a theory that is largely rejected by science but that Dial has resurrected.

Recapitulation theory is a notion that goes back to ancient Egypt, though it was formalized in the nineteenth century by the German biologist Ernst Haeckel. It holds that the early development of a single animal mirrors the evolutionary history of the species. Very young human embryos look like fish, for example, as the theory poses that our human ancestors did long ago. Dial doesn't agree that this is true all of the time, but he believes it sometimes appears to be true.

Students in a graduate seminar he was teaching in the late 1990s, Dial tells me, helped set him on his path to investigate how flight may have first developed. As part of their assignment, they studied the origin of flight and interviewed published researchers in both the arboreal and cursorial schools. They concluded that there wasn't a lot of good data for either theory. So at the end of the seminar, the students issued Dial a challenge. Why didn't he, the functional morphologist, do some research and come up with a new take on this question of the origin of flight? Dial thought that was a fine idea, since the study of the subject is indeed "very limited by the fact that the animal has been replaced by stone. It's not moving, just an anatomy left for us to try and interpret." The two theories, based on fossils and very little concrete evidence, "constitute a lot of arm waving and little data," according to Dial. "It's easy to come up with a hypothesis with a simple dead structure you want to fit into your story. That's just storytelling. My feeling was that we need to understand broadly and in depth the anatomy and physiology of the living." Dial believed he could use his high-speed cameras and other sophisticated instrumentation to come up with some new ideas about the origin of flight.

Dial is well suited to take on the subject from a new angle. He's observed hundreds of species of wild birds for decades and taught field courses on birds across Africa. As a researcher, he has long and meticulously studied the physiology of birds, every component part, muscle, nerve, and bone, and how this equipment determines how birds fly and run. As the host of the acclaimed series *All Bird TV* on the Discovery Channel, he has also traveled across North and Central America, filming birds in the field and interviewing a wide range of other bird experts.

It was in an unlikely source that Dial found a glimpse into the likely origins of flight: baby birds, who, in those first few weeks of their existence, he believes, provide a detailed look at the millions of years it took for the ability to fly to evolve. It's just one example of how modern-day birds inform our knowledge about the very distant past.

It's hard to fathom the notion that the feathery little creatures that flock to our feeders are modern-day dinosaurs, but it's true. All birds are, though chickens and turkeys are genetically the closest dinosaur relatives. Researchers, in fact, have manipulated chicken genetics so chicks grow dinosaurlike beaks, feet, and legs. It took a scientific revolution to overthrow the notion that dinosaurs were reptiles, a revolution that purportedly arose in 1868, when the biologist Thomas Henry Huxley, so fierce an adherent to the idea of natural selection that he was known as Darwin's bulldog, was eating dinner one evening while thinking about a dinosaur bone he had earlier been working with in the lab. As he nibbled on the bottom of a turkey drumstick, he was struck by its similarity to the anklebone of the dinosaur.

There was also a peculiar fossil find in 1861 in Germany that fueled such thinking: the discovery of the archaeopteryx. About the size of a raven, the strange creature had broad wings and long feathers and could fly, or at least glide, but actually had more in common with small dinosaurs than birds, including sharp teeth and a long bony tail. Archaeopteryx was a theropod—Greek for "beast feet"— a group that includes small to gigantic dinosaurs, including *Tyrannosaurus rex,* with feet similar to birds. Theropods had numerous other birdlike features: They brooded their eggs, had bones filled with air pockets to make them lighter, and in many cases had feathers and a wishbone, or furcula.

Birds and dinosaurs, Huxley thought, had to be related. It was a marked departure from the prevailing idea in the nineteenth century that birds were descended from a predinosaur reptile. It also implied that their ancestors had been warm-blooded, not cold. Huxley's idea got little traction, and the notion of a bird-dinosaur connection was largely forgotten. In the 1960s, though, Yale University's John Ostrom found twenty-two features in the skeletons of dinosaurs that were also found in birds, and he refloated the hypothesis. The "dinosaur wars" were underway, reaching a fever pitch in the 1980s as experts battled over whether dinosaurs were birdlike or reptilelike. By the 1990s, though, the matter was largely settled science.

Because they are related to dinosaurs, birds have been enlisted by scientists to tell us something about their ancestors. Dial turned to an unlikely subject—a baby chukar partridge. "I've worked in the lab with everything from pigeons and starlings, parakeets, magpies, finches, ducks, geese, and swans," he says, "but chukars fit the bill."

There are two forms in which birds are born into the world. One is *altricial,* which means "requires nourishment." Altricial birds, such as the robin, are feeble, naked, and helpless after they burst through their eggshell, and they need days of doting care from parents, who keep them warm and stuff food into their perpetually wide-open mouths as they develop feathers and the ability to fly. The other radically different way birds arrive is *precocial,* mind-bogglingly mature at birth, able to walk or run, evade predators, and forage for food on their own. Precocialism occurs along a spectrum and is especially pronounced in ground birds, who are most susceptible to predators. Some unusual birds, such as the megapodes—"large foot"—of Australia and New Zealand, are *superprecocial,* which means they are the most mature birds at birth of all. And Australian brush turkeys are the champions of superprecocial behavior.

Brush turkeys are bold, with no fear whatsoever of people. These days, they hang out in parks, near malls, and in other suburban settings. They look something like the wild turkeys found in North America—a deep blue-black color, with a bright head and flat tail. They forage through leaves and grass on the forest floor for seeds, fruit, mice, frogs, and other small animals, and they build unusual giant communal nests out of leaves and twigs that can be as large as twelve feet wide and six feet tall. They are known as incubator birds, because when the hens lay their customary eighteen to twenty-four large white eggs, the heat from a compost pile they build beneath the nest radiates up to keep the eggs warm. The males tend the eggs carefully, sticking their beaks into the compost to monitor its temperature and adjusting the mix to make it warmer or cool it down.

As soon as the babies peck their way out of the shell they are ready to roll. At one day old—one day!—the fully feathered chick's eyes are wide open, and it literally hits the ground running. With extremely powerful legs and feet, these little chicks are able to fly and climb vertically up trees and rocks, abilities that allow them to evade snakes, dingoes, and other predators. "The hatchlings are very comfortable on extraordinary slopes, with no panting or stress, doing it over and over again," says Dial. On that magical first day, a baby brush turkey can fly and run up steep and rocky cliffs better than an adult, though that ability eventually diminishes.

The chukar partridges that Dial studies are also precocial—though not superprecocial—ground birds. Chukar partridges are small and stocky, a member of the Galliform family, which also includes chickens and pheasants. They have a distinctive white face with a black gorget, or collar, of dark feathers. They favor dry, harsh terrain, and they eat a mixed diet, a buffet of insects, grass seed, roots, and whatever else they can scrounge. They were introduced to the United States from the deserts and mountains of their native

Pakistan because they are fast and flap furiously when they take flight, which makes them an appealing quarry for bird hunters.

Chukars, like pheasants and ostriches, lay their eggs in a depression in the ground. After a couple of weeks, the young hatch, and within twelve hours they run and start foraging for food, though they are still occasionally tended to by parents. There was something in the chukars, Dial thought, that was very dinosaurlike—it was the way they used their four appendages, something he calls "wing and leg cooperative use." They also maintained a wildness about them when he brought them into the lab. The Japanese quail he was using, on the other hand, which are commonly used as lab animals, were docile and sat quietly in his hand as he stroked them. Chukars could be raised and handled, yet they remained skittish and fled when he came near them—a perfect combination of traits, he says, for studying how they learn to fly as a window into how the dinosaurs might first have learned to take to the air.

Dial and his research partners placed a covey of downy baby chukar partridges on the floor and moved one of the babies away from the rest of the brood. Using the same high-speed cameras he and Bret Tobalske, director of the Flight Lab, used to study hummingbirds, they then filmed the chukar to study how the bird deployed its legs and stubby developing wings to instinctually and hastily scramble back to its brothers and sisters. As the bird got older, it was moved farther and farther away from its siblings. One day the rancher who was selling Dial the baby chukars came in to observe the research operation. "What the hell are they doing on the floor?" the rancher asked, shaking his head in dismay. "You call yourself a doctor? They like to be *off* the ground." This provided another piece of the puzzle—just as they would in the wild, the birds needed to be able to climb varying heights, on textured surfaces so they had traction. That could be a critical part of learning to fly. Dial brought some hay bales into the lab, replacing the slick boards he had been using. Each day he moved the bales a little farther apart, watching and filming the birds as they made their way between them, and placing foam padding on the floor in case they tumbled to the ground. The first day the hay bales were two inches apart, and within a week they were three yards apart. "We were trying to document on a daily basis how these animals develop performance in flying," says Dial.

It was Dial's fourteen-year-old son, Terry, who provided the most surprising insight to support Dial's theory that we can view bird babies as a proxy for dinosaurs. The young man was working for his dad in the Flight Lab in Missoula where Dial Senior had set up the bales. When Dial was back at Harvard, collaborating on research, he called to check in with his son, who was running the camera on the partridges in his absence. "How's it going?" he asked Terry. "It's going lousy," his son answered. "They aren't flying anymore, Dad. They're cheating."

"Cheating?" Dial asked. Yep, his son told him, they weren't trying to fly between the stacks of hay, but rather running straight up the side of the bale. Dial scratched his shaved head. How could they do that? No one had ever documented that type of climbing behavior in a bird.

It became a key point in his theory and the focus of his bird research: running up vertical surfaces was how birds had first developed their flying performance. He brought different species to the hay bales and put them through all their paces and, lo and behold, it turned out that a wide range of birds effortlessly ran straight up the sides of the hay bales.

Dial took other videos, of different bird species climbing trees, rocks, hay bales, and other inclines—babies, juveniles, and adults, including an owl that ran straight up the side of a tree trunk. And it was while watching this unusual behavior that "it dawned on me," Dial said, "why it is that birds have half a wing."

Baby chukars flap their small, developing wing stubs as they run, he explained, in a way that improves the traction of their legs, or hind limbs. It allows them to climb up a steep slope, whether chasing a grasshopper or fleeing a fox. Animals strive mightily to find an elevated retreat for safety when they are fleeing, but if the slopes are more than sixty degrees they fall backward. A developing proto-wing, however, acts like a spoiler on a race car, and combined with powerful legs and feet, it keeps the bird firmly on the surface, allow-ing it to climb the most vertical slopes, and even to walk upside down on an overhang. No matter what the bird's body is doing, the bird keeps the wing at an angle of about twenty degrees, and the wing both powers its running and provides stability. Then, as the bird grows up and begins to fly, it uses the same wing angle. Dial believes that what he calls wing-assisted incline running, or WAIR, is a fundamental and dramatic bird behavior that, until now, natu-ralists and ornithologists have missed.

A half wing, then, was not a useless appendage, as Mivart and a host of others believed, but instead essential, because it turbo-charged the climbing abilities of young dinos, increasing their chances of survival. Just like the baby chukar, precocial or superpre-cocial dinosaurs likely used their wings to power their running—away from predators or toward a fleeing meal—before they evolved the ability to fly. "If you are born with very little parental care, a kid has to do what it's got to do, or it's dinner," Dial says.

Moreover, his son Terry, now a Ph.D. student at Brown in life his-tory biology, studied baby ducks as they learned to fly. It takes a long time for duck wings to grow to maturity, because they need to be especially strong for the birds' long-distance migrations. While those wings are forming, the elder Dial says, babies still need to be safe from predators, so the growing wings power the ducks in their swim-ming to escape danger, in the same way that they help chukars run away.

As Ken Dial's partridges grew, not only did they climb the bales with their feet and wings, they started flying in the lab, flapping and gliding from the tops of the hay bales to the linoleum floor. This is what Dial thinks likely happened with dinosaurs as they began the early stages of flight. Early wings first powered the precocial dino-saurs' running and climbing, and much later—perhaps millions of years later—as they climbed rocks and flapped back to the ground, these developed into flying wings, and those dinosaurs with the bigger and better wings could more effectively escape predators. Feathers were vital, too, as those dinosaurs with the biggest and lightest feathers—the best for escaping predators, for soaring, and for finding food—also survived.

Dial gives this hypothesis the less than catchy name "ontogenetic transitional wing hypothesis," or OTWH. And it has been well-received in the field.

One of those who finds some common ground with this theory is Richard Prum, an ornithologist at the Yale Peabody Museum of Natural History who is in the arboreal, or flying-out-of-trees, camp. "Lift is always cheaper at speed," he told me. "The aerodynamic force that makes flight possible is that air moves at different speeds over the top and bottom of the wing. Those forces are greater the faster the air is moving. Gliding animals—draco lizards, flying squirrels—plummet, and at the end of the plummet you have finally gone fast enough and you can really move. That means more lift. So, it's much more likely that at the first moment at which lift was created with a feather, it was probably because of falling out of a tree, falling from a height."

Dial's theory, in a sense, provides support for both the ground-up and trees-down schools of thought. It lays out a plausible way for the first flying prehistoric animals to have flapped their way up into the trees and glided down, as aborealists believe. But he thinks both camps are wrong. In today's world there are absolutely no examples of gliding animals, anywhere, says Dial, that flap their flaps to stay airborne. So he doesn't think that gliders became fliers back then, either. His studies also demonstrate that birds have strong running abilities, which supports the cursorialist argument. Dial, however, doesn't believe the first fliers ran until they went airborne, again because there are no contemporary examples of it. "No bird today runs to go airborne," he says. "Nowhere in the world. But birds flap up inclines and flap back to the ground. So the behavior I offer for the first fliers"—that is, birds climbing up a rock or a tree and flapping down—"is seen everywhere in all birds."

While wings get most of the attention when it comes to birds, Dial argues that the most overlooked part of their miraculous anatomy is their feet, even if they are not exactly elegant in appearance. "Bird feet are spectacular, a multifunctional tool like a Leatherman," Dial says, after we watched the bird-climbing videos. "A bird foot can do whatever a bird wants it to."

Dial describes to me three types of walking. He holds up the first finger to further explain. "You and I and bears are *plantigrade*. In other words, we plant our heels and toes and walk off our toes." He holds up a second finger. "Your cat and dog are *digitigrades*. They walk on their digits. Lastly a horse, wildebeest, or goat walks on its toenails. It's called *unguligrade*," he adds, raising a third finger. "Birds," he concludes, with admiration, "do all of those things."

"My story is this," Dial continues enthusiastically. "Look at this animal. The day it hatches it uses all four limbs to climb like a lizard up to safety. And six days later, though it can't fly, the baby uses its wings to control descent and land perfectly. A week or so later it flaps down, getting better and better. And a week later it can get vertical and fly. I would argue that these stages are, metaphorically, how flight evolved. But while it may take thirty days for a bird, evolution probably took many millions of years."

To paleontologists and others who traffic in information about the deep past, Dial advises, "Study warm-blooded living dinosaurs, not only their cold stone remains. Let's go with what the most compelling data are, and not just who can wave their arms around the best, and perhaps articulate their argument best. The best way to do that is to study the living. Don't get me wrong, I am all for studying the past with fossils. I get down on my knees and thank God for paleontologists who are digging in the dirt. But everybody is caught up with the dinosaurs, because they see the tyrannosaurus or the velociraptor and they are sexy, while birds are tweety little annoyances. The researcher who studies birds is a bird nerd, and the other is a macho paleontologist. But birds are the dinosaurs that made it because they have been able to change with the times. That's what makes them so extraordinary." Birds can provide many other glimpses into the era of the dinosaurs, he says. From birds we can gather information on how dinosaur parents fed and nurtured their young, for example, or what their habitat was like. "This idea also says to the birders, look at the glorious information in the babies. Stop being so consumed with the adult form. They are cool, but look at the babies, they are really, really cool. They are telling us the most about evolution, about ecology, and form and function, because they are changing every single day and yet have to interface with their environment every day as they change and survive."

And though every day we see birds flying around our house, building a nest in the garage, or grabbing worms in our yard, we never suspect that those critters are the living, breathing descendants of the mysterious giants that walked the earth more than a hundred million years ago, and that they provide us a glimpse into that long-ago world.

⌒〰⌒

# CHAPTER 2

# Hummingbirds:
# The Magic of Flight

> If I had to choose, I would rather have birds than
> airplanes.
>
> —CHARLES LINDBERGH

Why have people across the ages longed to leave the ground and soar? Birds and bird flight have stirred us at fundamental levels across the span of human existence, eliciting feelings of elevation and transcendence. Many of us ache to let go and leave our earth-bound life behind, physically and emotionally, and birds can live out this dream for us. With an unimaginable lightness of being, their flight is universally stirring; the pleasure it offers the observer cuts to the core of one's being.

One of the first recorded human attempts to fly like a bird was that of the ninth-century Spanish Muslim polymath Abbas Ibn Firnas. At the ripe age of sixty-five, Ibn Firnas, inventor of such devices as a water clock and vision-corrective lenses he called "reading stones," built a primitive wing suit, a wooden-framed contraption covered with vulture feathers that he fixed to his shoulders and arms. With a crowd looking on, he leaped off a high wall wearing this crude flying device. Accounts are varied, but most say it kept him airborne momentarily, and then he came down—hard. "His back was very much hurt," wrote one historian, "for not knowing that birds when they alight come down upon their tails, he forgot to provide himself with one."

For his part, the eleventh-century "flying monk" Eilmer of the Malmesbury Abbey, in Wiltshire, England, was captivated by the story of mythical Daedalus, who fashioned wings for himself and his son Icarus out of wax and bird feathers. Icarus flew too close to the sun, melting the wax that bound the feathers on his wings, and so he fell to his death. Inspired also by a growing realization among thinkers at the time that bird flight was not magical but based on physical principles, the young monk fastened homemade wings to his hands and feet—after watching jackdaws fly—and leaped off the top of the abbey. He descended one hundred fifty feet and covered some two hundred yards before crashing into a swamp, where he "broke both his legs and was lame ever after," according to one twelfth-century historian. "He used to relate as the cause of his failure, his forgetting to provide himself a tail."

Leonardo da Vinci, too, had a fever to fly and would do much in his life to capture the essence of bird flight, but as an artist and scientist he took a much more studied approach than Ibn Firnas or Eilmer. Leonardo would sit on a hilltop near his home in Florence and sketch the black kites and goldfinches that soared and flitted above him. These sketches eventually became *Codex on the Flight of Birds,* the first known written record of the study of avian aerodynamics. The codex is filled with five hundred or so elegant black-and-white pencil sketches of birds in flight, studies of their wings, and drawings of flying machines modeled after them.

Leonardo's observations about bird flight were remarkably prescient in the 1400s, centuries ahead of other scientists. So little was known about birds at the time that many people believed that birds turned to stone and sank to the bottom of lakes to hibernate or went into a torpor or even migrated to the moon, instead of flying to warmer temperatures in the winter, because there was no way to know where they went. It wasn't until nearly the turn of the eighteenth century, with the publication of Thomas Bewick's *A History of British Birds,* that it began to be accepted that swallows and other birds migrate. And in 1822, the principle of migration was illustrated dramatically when a white stork was shot in Germany. When the hunter fetched the bird, he found that it had a spear in its neck—an African spear. It had flown, wounded, from Africa to Europe.

Leonardo noticed the birds' distinctive shape, how they moved through the air, and he observed that air behaves much like water, which is key to understanding how birds fly. He noted the flow of the air over and around a flying bird and how it exerts lift beneath the wings and tail. He realized that the pressure on top of the wing is less than the lift that pushes up on it from the bottom, which is the fundamental aerodynamic principle that keeps birds, and airplanes, aloft. And he recognized the importance of the tail. "He went straight to the heart of how birds fly," says Prum, the ornithologist at Yale University. "It's fantastic, almost eerie, how much progress he made."

A genius inventor, Leonardo even built some of the flying machines he sketched, the most famous of which was the ornithopter, a contraption with human-powered mechanical wooden wings that mimicked the up-and-down flapping of birds in flight. He launched them from a hill near Florence, but they failed to make it aloft.

Humankind's fervent wish to defy gravity would not be fulfilled until more than three centuries later, when real flight—twelve seconds' worth—was achieved by Orville and Wilbur Wright, who had studied Leonardo's codex (and his writings on birds) for hints on designing their flying machine. While watching buzzards soar along the Great Miami River near his home in Ohio, Wilbur realized that the birds maintained their position in the air by making slight adjustments to the angle and position of their unflapping wings. As a result, when the brothers designed their first plane, it had gliding wings that could be minimally adjusted with pulleys and cables. With this invention, humans realized that they were no longer bound to the earth.

While research on bird flight has been going on for hundreds of years, researchers are just beginning to glimpse the full extent of birds' aerial capabilities. "Birds can do some pretty spectacular things," says Ken Dial, who has studied bird flight for more than thirty years. "They can go from forty miles an hour to zero and land on a branch that's moving, all in a couple of seconds. It's inspiring. And we still haven't come anywhere close to that."

The most spectacular flier of all by far is the hummingbird. All of the world's 350 species of hummingbird dwell in the Americas. They are a study of power in miniature. They range in size from two grams—the weight of two paper clips—for the calliope to twenty grams, the weight of the giant hummingbird in South America. The eggs of the average hummingbird, laid in a nest the size of a teacup, are smaller than kidney beans. Hummers have the broadest array of colors of any species of bird, and their feathers seem to glow from within. That's because instead of a pigment in their feathers, as most birds have, their plumage is comprised of crystal-like cells that gather and reflect light, like tiny glass beads. The first European explorers to encounter the vividly colored birds whirring among the emerald green jungles of South America were entranced, and referred to them as *joyas voladoras,* flying jewels.

For such a tiny, nearly weightless bird, hummingbirds are remarkably robust, often migrating extremely long distances in deeply adverse conditions. Calliopes fly nearly three thousand miles annually from the northern United States to Mexico, at speeds up to thirty miles an hour, dodging hurricane-force headwinds and wind-driven raindrops nearly as big as they are. They are the first birds to return

to the northern states each spring, in late March or April, long before any flowers have bloomed, when they survive on spiders, mites,
and other insects in the snow-shrouded landscape. In the Andes they
are a symbol of resurrection, because on cold nights, they go into a
state of torpor, dialing down their physiology to the point where
they appear to be dead. Then, in a seeming miracle, when the sun
warms the forest in the morning, they spring back to life and start
buzzing among the flowers.

The heart that powers this flying machine is the size of a pea, yet it
is the largest in the animal world relative to the animal's size. It beats
hundreds of times per minute—the record number of heartbeats is
a whopping 1,260 times a minute, measured in a blue-throated hummingbird. In torpor, hummingbirds' heart rate oscillates between
fifty and two hundred beats per minute.

The little birds are incomparably and exquisitely light, fast, and
maneuverable, able to fly upside down and backward, and, as nectar-
ivores that gather as much as twelve times their own body weight in
sugary nectar each day, they are masters of the sustained hover. They
execute these midair refuelings more expertly than any aircraft,
floating effortlessly in front of a flower and drinking its nectar without spilling a drop, even as they are buffeted by stiff breezes or driving rain.

To accomplish these flying feats, hummingbirds beat their wings
as much as ninety to one hundred times per second, even more when
they are in the thrall of courtship rituals. In contrast, a common pigeon beats its wings about nine or ten times per second in flight.
Hummingbirds flap so fast that as their wings are starting to go up,
their brain is already generating the electrical signal for the downstroke. Bigger birds have more overall power, but the rapid speed of
the hummer's strokes makes the bird more powerful for its size than
any other. "It's the difference between a sports car and a truck," says
Dial.

Hummingbirds have evolved specialized adaptations to make
such virtuosic flight possible. Like all birds, they have pneumatic

bone, which is either hollow or filled with air sacs, and instead of teeth, they have a very lightweight keratin bill. The brain's cerebellum, which gives birds the excellent fine motor skills necessary to choreograph their complex ballet of flying maneuvers, is large for the hummingbird's body size.

While the pectorals, or breast muscles, are the largest part of all birds' anatomy—about 80 percent of their weight—which enables them to generate the tremendous power needed for flying, hummingbirds have, proportionally, the largest pectorals of all. And the little creature has another difference: While most birds flap with the equivalent of their whole arm, the hummingbird uses just the wing portion near its wrist, which allows it to create lift with its wings on the downstroke, as other birds do, but then, by turning the wing over, it also makes the most of the upstroke. Hummers have also evolved an unusually large number of mitochondria in their cells, which are essential to the engine that powers bird muscles. "They are like carburetors," says Bret Tobalske, who studies bird flight with Dial, "and they have packed more into their tissue than any other bird."

The Flight Laboratory that Dial designed is one of the top institutions in the world for the study of animal locomotion. Headquartered in a remodeled nineteenth-century Mission-style cavalry barn, which is partitioned into more than a dozen rooms on two floors, it's a warren of offices, a warehouselike laboratory full of electronic equipment, X-ray machines, laser equipment, two wind tunnels, and a small surgical room where birds can be implanted with devices to measure such things as their muscle performance and oxygen intake.

Tall, thin, and athletic, with dark hair and a thick, dark beard, the soft-spoken Tobalske lives and breathes flight. He is an expert in the woodpecker's novel type of "intermittent bounding flight"—the way the bird flaps rapidly and then tucks its wings to glide for a while. He has put rhinoceros beetles through their flying paces. Now he is happily focused on hummingbirds. "No other bird comes *close* to being able to fly like a hummingbird," Tobalske says. "Their default setting is to fly. Whereas every other bird prefers to sit or perch as a reward for performing in the lab, hummingbirds are more energetic and they simply fly all the time and don't seem interested in resting"—a great attribute for those who study flight.

Their personality is another reason to work with the birds. Tobalske describes the hummingbird's "incredible warmth and tractability," which means they happily perform tasks on demand. "They've got that mammal brain," he says. "They are very bright. We hold sugar water away from them and they learn, just like a dog or cat would, that pretty soon the food is going to be available, and they do just what we want them to do."

On one visit I watched an experiment with a calliope hummingbird, which Tobalske had trapped in his backyard. As the Rolling Stones played quietly on a boom box, Tobalske gently cupped the iridescent red and brown bird in his hand and then released it into a three-by-three-foot clear Plexiglas box that had been filled with a foglike mist made of extra-virgin olive oil. The newly caught bird seemed alarmed at first and tried to fly up, repeatedly bumping into the ceiling. Meanwhile, Tobalske aligned a green laser beam to shoot through the Plexiglas and strike an area in front of a feeding tube filled with sugar water, the bird's favorite food. When the tube was uncapped and the bird zipped over to feed, it would be illuminated by the pulsing green beam.

The hungry hummer quickly found the tiny glass feeding tube, plugged its beak into the nectar, and hovered, drinking in the sweet water. Tobalske laughed in amazement as he watched the bird do its thing over and over. "Just try to get an insect to do that!" he said. "You can't. An insect just responds to chemical cues; they don't respond like an animal. Hummingbirds are like robots. They'll return again and again to the feeder. You can't do that with any of the ten thousand other species of birds."

The laser was switched on, and for about four or five seconds the bird was enveloped in flashes of emerald green lightning. It continued to feed unperturbed. All the while, a special high-speed camera that shoots one thousand frames per second—the same type the military uses to film ordnance explosions to study blast physics—recorded the flight. Tobalske and his team of graduate students would later watch the films to distinguish the bird's tiniest, fastest movements, far too fast to be seen with the naked eye. Leonardo da Vinci had to imagine the forces at work around the kites and finches he watched, but thanks to the laser-illuminated mist of atomized olive oil, all here is revealed. The incandescent mist travels around the hummer's wings and body as the air would, and when captured by the camera, it clearly demonstrates the complex forces constantly at work as the bird flies.

Later Tobalske placed the calliope in a wind tunnel—the equivalent of a bird treadmill. At one end was a feeding tube, and beyond that a giant fan. The bird headed straight for the feeder, and as the speed of the fan was slowly cranked up to a stiff twenty-mile-per-hour headwind, the hummingbird, intensely focused on the sweet juice, effortlessly ramped up its speed to keep its beak snugly in the tube. As a result, the bird was able to fly in place in a windstorm as the high-speed camera recorded its movements. "They are off the charts in terms of what they can do," Tobalske said as the bird's wings whirred away. "A hummingbird can hover like a helicopter for one and a half hours, nonstop, and no other bird can do that."

Hummingbirds at the Flight Lab are also subjected to an X-ray machine. Tobalske buys platinum jeweler "splatter," tiny specks of metal left over from soldering, and glues them to muscles on the bird beneath the feathers. The X-rays pick up that metal as the bird flies. The scientists can see the skeleton through the X-ray and can understand the loads placed on the delicate bones, how they bend and support the force generated, and the tug of gravity the birds can endure during flight. Once they have flown their way through the experimental gauntlet, Tobalske releases the birds back into the wild.

What have researchers learned with their battery of studies to add to what Leonardo and the Wright brothers knew? "That birds misled humans on what it took to fly, for a long time," says Dial.

"What appears to be simple and straightforward is anything but. The power and coordination of birds are far too advanced for humans to copy. A lot of people jumped off cliffs with wings and killed themselves, trying too hard to mimic birds. The act of flying is highly complex, is inherently unstable, and takes tremendous energy that we don't have as humans. It wasn't until people abandoned flapping flight and behaved more like a glider, like a soaring vulture as the Wrights did at Kitty Hawk, that we could do something useful.

"We've come a long way in understanding how birds fly since Leonardo da Vinci," he says. "But we still have a long way to go."

One thing is for sure, as the flying monk and flying Muslim both realized: Tails are very important. "A tail confers stability and aids maneuvering," says Tobalske. "Tails are integral to coordinated flight." One research hummingbird that had lost its tail is a case in point. Released to fly in a wind tunnel, it immediately corkscrewed to the floor. After a couple of attempts, though, it recovered and learned how to compensate without one, an act requiring far more strength and energy.

Birds also frequently change their shape as they fly, a technique that will likely someday be a major element of advanced human-powered flight. "Birds are constantly morphing, and morphing on different levels," Dial says. "The birds do this so eloquently, so elaborately, it's humbling. With every wing beat they go from the shape of a bullet to the shape of a kite."

A number of planes over the years have been designed to copy birds' ability to morph, though the scope of the morphing has been extremely limited. Landing and takeoff flaps are modestly used these days, while more ambitious morphing technologies to increase flight performance have proven difficult to perfect. But recently NASA and MIT built and successfully tested a shape-shifting wing, made from millions of pieces of metal, plastic, and other materials, that morphs nimbly very much like a bird wing as it flies. More testing is needed, but it may lead to the next generation of aircraft.

Biomimeticists are using other bird flight ideas. The feathered ridge design on the leading edge of an owl's wing, which renders it silent as a specter as it hunts, is being adopted to quiet airplane engines. The tweak of adding an upturned "winglet" at the end of a passenger plane's wing—which soaring birds like eagles and buzzards have—saves forty-five thousand gallons of fuel per jet plane annually. And Airbus, the European aircraft manufacturer, just refashioned their airplane wings to function much more like bird wings, a redesign that is expected to save enormous amounts of fuel and reduce $CO_2$ emissions.

Planes are not the only beneficiaries. When the West Japan Railway Company's Shinkansen bullet train was first built, it emerged from tunnels at two hundred miles per hour, causing loud sonic-boom-like sounds that brought complaints from neighbors. The railway's chief engineer, Hideo Shima, a birder, wondered what creature in nature traveled between two very different mediums. After much deliberation, the front of the train was modeled after the long, slim beak of the kingfisher, a fish-eating bird that dives from a perch above the river and, because of its aerodynamic beak, cleaves the water with very little splash. Not only did the bird-inspired shape hush the train, but the sleek design increased fuel efficiency, reducing electricity consumption by 15 percent and upping the speed by 10 percent.

It will likely be a very long time, if ever, before a human aircraft flies like a bird—especially a tiny superbird like the hummer. "Humans will never be able to generate enough power to carry their weight in the same ways, which is what the hummingbird has mastered," says Tobalske.

The flight of the hummers has inspired a generation of small, highly maneuverable robotic craft. A team of engineers has created a modern version of Leonardo's ornithopter, called Fullwing, that borrows the hummingbird's extended wing-flapping approach. And defense researchers are building a nanohummingbird, a remotely piloted flying robot that looks like the little bird and crudely approximates its flight performance, zipping down streets and through doorways and carrying a camera in its neck to spy on evildoers. It's eyed as a tool for urban warfare.

Before Tobalske places his hardworking calliope back into its cage, he asks if I want to hold it. He sets it down on my palm, which I gently close. It is the first time a hummer has graced my hand, and I worry I will crush the little guy. It moves frantically for a minute and then calms. I am struck by its remarkable lightness, little more than an idea, it seems, with a rapidly thrumming heart I can sense.

I realize as I watch the bird's extraordinary acrobatics and feel its tiny, trembling heart that birds—not just hummingbirds, but all members of the bird kingdom—embody the most appealing aspects of nature. "Few other creatures seem so alive in every fiber," wrote the poet Saint-John Perse. "Even in quiescence they are concentrations of vitality." They are our daily connection to nature's miracle, enchanting us with their exquisite and extreme colors, songs, shapes, sizes, and abilities. They call to us from another time, evoke deep memories of our evolutionary past, and remind us of our history in the wild. And of course, they represent our deep longing to fly. "Once you have tasted flight," wrote Leonardo, "you will forever walk the earth with your eyes turned skyward, for there you have been, and there you will always long to return."

# CHAPTER 3

Canaries and Black-backed
Woodpeckers: Birds as Flying Sentinels

The woodpecker's drilling
Echoes
To the mountain clouds.

—DAKOTSU IIDA,
JAPANESE HAIKU POET

The Canary Islands are an archipelago of seven islands off the coast of Morocco, their name taken from the Latin for "Island of the Dogs," Insula Canaria, after the wild canines that roamed the tropical landscape. When the Spanish conquered the islands in the fifteenth century they heard the sweet trill of tiny, colorful birds—which they named after the islands—flitting among the jungle branches and saw the potential to breed them as pets. Within years, a canary trade flourished across Europe. Since only male canaries sing, entrepreneurs kept a tight rein on the supply by selling just the males. No one else could breed them. That changed in the 1600s, legend has it, when a Spanish merchant ship went aground near Italy and its cargo of canaries flew free. Locals captured the birds and began breeding them, which broke the hold of the Spanish monopoly.

A pair of canaries were routinely packed as safety equipment into the nineteenth- and twentieth-century coal mines of Wales and England, and later U.S. miners took them to work. As they swung a pickax or loaded ore cars, workers kept one wary eye on the little yellow birds. If they saw one lying distressed or dead in the bottom of the cage, they knew carbon monoxide levels had reached a dangerous level and they could well be next, so they scrambled out of the mine. That's how the little bird with the sweet voice, named after wild dogs, became one of the most familiar metaphors in history—the canary in the coal mine.

The use of the little birds as sentinels led to a one-of-a-kind invention—a resuscitation cage for mine canaries, a square metal mesh-fronted box with a tiny oxygen tank on top. If the canary, wigged out on methane, plummeted to the floor of the cage, the box could be sealed and the oxygen turned on until the little bird came to. This way, a canary could be used again and again. Canaries served as toxic gas monitors until the 1980s, when they were phased out and replaced with electronic equipment. But the role they pioneered for birds as sentinels for humans lives on in big ways.

We might shudder at the destruction caused by a wind-whipped fire reducing a verdant forest to a pile of ashes and leaving behind standing trees that look like giant burned matches, but the Montana ecologist Richard Hutto sees things very differently. Perhaps only pyromaniacs and ecologists think this way, but as long as homes or human lives aren't at risk, Hutto is downright thrilled at the prospect of a big blaze. He explained his thinking to me late on a July morning as we walked along a trail through the scene of a decade-old forest fire on Black Mountain, near the southern edge of Missoula, Montana. The day had already warmed and unlocked the fragrant smell of pine, and dragonflies buzzed lazily about. Hutto carried a flashlight-size digital recorder, and as he walked he played a recording of the drumming of the black-backed woodpecker. Each woodpecker species has its signature tattoo, and in the case of the black-backed it's a long drum that starts out slow and picks up speed.

Hutto is another bird nut turned pro, his life devoted to the study of birds as a professor of ornithology at the University of Montana. In his sixties, he looks considerably younger, with freckles and reddish-gray hair and the energy of a fortysomething who rapidly hikes long and arduous distances in the field. As we walk, without prompting he spontaneously and capably—and often—imitates a wide range of birdcalls, and his cellphone ring tone chirps with a birdsong, too. "That's the olive-sided flycatcher," he says as he answers it. His specialty is deciphering bird numbers and behavior as a sign of what's going on in the natural world, a way of detecting and measuring ecological changes.

That's why we are searching for the woodpeckers. We walk through the woods for an hour, frequently playing the drumming, though our calls aren't returned. Just as we are about to leave the forest, the bird's distinct tattoo echoes through the woods. "That's it," he says, and scribbles some notes.

Wildfire is as critical to the forests of the western states as rain and snow. Most trees here don't grow old enough to fall to the ground and decompose into soil as they do elsewhere. Instead, a hot flash of wildfire reduces forests to ash, which unleashes high levels of nitrogen and phosphorus, which in turn spark the growth of a new forest. For thousands of years, wild blazes have periodically burned through the vast blankets of ponderosa pine and lodgepole pine forest in the West, from small fires that slowly creep across the ground and burn grass and small trees to full-blown conflagrations that blacken square miles. The nine-inch-tall black-backed woodpecker evolved on these crispy landscapes, along with a suite of other charred-forest "specialists" from plants to insects. While the ashes are still warm, these species rush in.

Even while a forest fire is still smoldering, the jewel and fire-chaser beetles are zooming toward it. With infrared sensors under their spindly legs, these bugs are highly sensitive and can detect heat from forty miles away or more. When cigarette smoking was common, they would swarm crowded baseball stadiums because of the heat and smoke from so many lit cigarettes. When they reach the fire's aftermath, they stake a claim to a patch of burned tree, which, now dead, is defenseless. The black-backed woodpecker soon follows, and it feeds on the grubs that hatch from the eggs left behind in the charred trees by the heat-seeking bugs. The woodpecker has a white breast and a yellow slash on its crown, but it has evolved a coal-colored back and wings to help it blend in with the charcoal trees as it feeds, so as not to be plucked from its perch by a hawk or falcon. Hutto says he has never seen these woodpeckers outside of burned areas, and he knows of no other species of bird so dependent on one type of habitat. That's saying something, because since the 1970s Hutto has crisscrossed both burned and unburned forests in the northern Rockies, playing his recorded call and scouring the woods for black-backed woodpeckers.

Even so, it isn't the black-backed woodpecker itself that Hutto is really interested in. Although burned forests that remain undisturbed by logging are critical to woodpeckers, they are also essential to many other species. A forest can grow slowly for decades or more without much serious change, and then a big fire wallops it and hits a reset button; it's a shot of adrenaline that results in an explosion of change, the beginning of a wildly dynamic new ecosystem. "After a big fire is when the magic happens," says a grinning Hutto. It's an ancient rhythm.

If there are not enough severely burned areas because fires have been prevented or minimized, or if the burned trees are cut down and sold for timber, a full ecological transformation is thwarted. Many insects disappear, for example, because the trees are gone.

That means woodpeckers will not find enough to eat, and their numbers will decline. The black-backed woodpecker, then, to Hutto's way of thinking, is an easily observable indicator of whether the postfire magic is happening—whether many other species that thrive in burned forests are doing well, birds such as mountain bluebirds and wood pewee, as well as tiny capped morel mushrooms and plants like the wild geranium and snow brush, whose seeds may lie dormant in the soil for a hundred years until a hot-burning fire wakes them up. Instead of studying all of these species separately to gauge the health of the forest, which would be nigh impossible, he simply monitors the number of black-backs. If their numbers are steady or increasing, there is plenty of burned forest to keep woodpeckers and everybody else healthy. If their numbers are trending down, Hutto takes notice; it means something is amiss. "Forest ecology is a complicated matter," Hutto says, "but a bird tells you in a very simple way that everything is okay from Canada to Mexico. They have something to say if we listen, and that's what I do, I listen to the stories the birds are telling us. In this case, they are saying that if humans don't appreciate the value of severe events, we may lose a lot of the plant and animal diversity stimulated by severe fire." It seems we aren't protecting enough of the severely burned forests in California, Oregon, and elsewhere where these woodpeckers live. While they aren't in danger of vanishing from the planet, their diminishing numbers reflect the decline of ecological integrity in our forests.

The black-backed woodpecker, then, is a metaphorical canary in a coal mine. Because they travel far and wide, are found everywhere, and are conspicuous, birds have become a widely used sentinel these days, the perfect creatures to reveal any number of ecological problems and changes in the world around us. "Name another animal that tells you it's there, another set of organisms that are as easy to monitor," Hutto says enthusiastically, referring to the way birds announce their presence with their song, a loud call, or an echoing, rhythmic tattoo. "I can stand here blindfolded," he says, waving his arm toward the broad sweep of forest, "and with ninety to ninety-five percent accuracy tell you every species of bird that is around me within the length of a football field. They are the first thing we see when we walk out the door, and they identify themselves. Nothing else does that. Holy cow, birds are phenomenal!"

Moreover, their charisma and beauty, he says, attract millions of bird-watchers to happily keep track of birds with great and meticulous passion—for free. "If people don't see a warbler in their backyard that they've been seeing, they'll let you know, believe me," he says. "If sixteen people call, you know something's up and you can investigate." It might mean that their habitat has been destroyed or that the birds have been infected by disease or something else.

One of the most famous examples of sentinel birds can be found in Rachel Carson's paradigm-shifting 1962 book *Silent Spring,* which documented in excruciating detail the impacts of pesticides on birds, and on human health as well. Carson, who died from breast cancer not long after the book came out, described a scene on the campus of Michigan State University after DDT was sprayed on elm trees for pests. "Dead and dying birds began to appear on the campus," she wrote. "Few birds were seen in their normal foraging activities or assembling in their usual roosts. Few nests were built; few young appeared. . . . The sprayed area had become a lethal trap in which each wave of migrating robins would be eliminated in about a week. Then new arrivals would come in, only to add to the numbers of doomed birds seen on the campus in the agonized tremors that precede death." DDT was later banned, largely as a result of her book.

These days, birds are routinely used to detect a wide range of dangers, from pathogens to pollutants. Hundreds of thousands of wild birds, both dead and living, are routinely tested for the avian flu, to get ahead of the next pandemic, and a range of wild birds are monitored as an advance warning system for Lyme disease. Caged chickens are set out in parks and other places, and their blood is tested to see if they turn up positive for West Nile virus. Seabirds are at the top of the food chain, and they accumulate toxins and so are indicators of the level of mercury, cadmium, lead, and other toxic substances in the environment.

New technologies have amplified the role of these flying sentinels. One of the first examples comes from the 1990s, when Brian Woodbridge, a Forest Service researcher in far northern California, encountered a Swainson's hawk mystery. He noticed that as winter came on, many of the hawks he studied—also known as grasshopper hawks because they eat the big bugs—were leaving Butte Valley, near Chico, California, in autumn and not returning come spring. No one knew where they went or what was happening. Birds were banded in those days, but the odds of finding them at the other end were often remote. Woodbridge heard about a lightweight satellite transmitter that could be fixed to the bird's feathers and would signal its whereabouts to an orbiting weather satellite.

In the fall of 1997, Woodbridge trapped two birds and fastened the transmitter, a little heavier than a silver dollar, to their tail feathers. They circled into the sky wearing the $3,000 instruments and headed due south, chasing summer. One disappeared, but two months later the other bird beamed a signal from a region in Argentina called La Pampa, six thousand miles from California. Incredibly, it was the first time anyone knew where the birds went for the winter.

To get to the bottom of the mystery of the disappearing hawks, Woodbridge and two colleagues traveled to the birds' wintering ground in Argentina the following year. They were astonished. After seeing birds occasionally in Butte Valley, he discovered that here the Swainson's hawks roosted in huge flocks—sometimes thousands of birds—in vast eucalyptus groves called *montes*. But something was obviously very wrong. As he drove to the ranch to find the bird with the transmitter, he passed hundreds of dead birds on the ground.

After a little detective work, Woodbridge figured out what was happening. The farmers in the area had just started using a new, deadly pesticide called monocrotophos. "The grasshoppers are building up, and they get to a point where they are attracting hawks, and the farmer calls a plane to spray his field," Woodbridge told me. "The hawks were diving right into the field in front of and behind the spray rig, catching grasshoppers coated with this compound." Death struck like lightning. Some birds perished with a grasshopper in their claws, having absorbed it through their feet and dying before they could eat it. In some cases, a fifth of the birds that roosted in a *monte* were killed by poisoned bugs.

It was a tragedy not only for the hawks, but for the farmers as well. A grasshopper hawk eats a couple of hundred hoppers a day, so the farmers were unwittingly eliminating the bird that was helping them. Woodbridge and others arranged a meeting with Argentine farmers and pesticide manufacturers, who agreed to quickly phase out the pesticides and buy back what hadn't yet been used. A new era of technology-assisted sentinel birds was just beginning.

John Fitzpatrick, known as Fitz, is the director of the Cornell Lab of Ornithology. Harvard-educated, he is an accomplished field biologist who studies the family dynamics of scrub jays. His job is to make a case for the protection of all of the natural world to the public; because of their charisma, birds, he says, are the best ambassadors. With white hair, a white mustache, and glasses, Fitz is an owlish-looking fellow, which seems appropriate. As we chatted in his office in front of a giant picture window that overlooks a large pond and a bird-filled forest, it became clear that one of his favorite topics is what birds tell us in great detail about the rest of the world, and how their sentinel role is best tapped. A pinnacle of his tenure is something called eBird, a smartphone app that aggregates the observations of millions of bird-watchers "to make the most detailed map of an organism ever made," he says, and to tell us things about the world people haven't dreamed yet.

Fitz lined me up to go eBirding with a devoted bird-watcher and retired state wildlife biologist named Bob Martinka near my home in Helena, Montana. On a sunny fall day, Martinka and I drove out to a small lake, and the biologist trained his spotting scope on a towering tree, where a dozen graceful, long-legged great blue herons sat in or near their nests. Martinka got out his smartphone and quickly typed in the number of birds and the species, then moved down the road to the next species. Martinka volunteers his services a few days each week, scanning lakes, prairie, and even the local dump, where, between rows of old refrigerators and a compost pile, he often sees rare gulls on top of a giant mound of bird-covered trash. Long a bird-watcher, Martinka is now also a "biological sensor," and there are tens of thousands like him all around the globe turning their bird sightings into data. His report of a dozen herons at the lake is just one small piece of information, but when it is sent off to the supercomputers at Cornell and combined with other sightings, it's like a tiny pixel that together with others provides a very big picture of what birds around the world are doing and in what numbers. In a sense, Martinka and his fellow sensors are like a collective mind that's bigger and smarter than its component parts. It's the first time scientists have had such a view, and it's changing the way they see the bird world.

One eBird revelation is a "heat map," a visually striking image based on thousands of gathered bird sightings that shows the movement of an entire species across the United States in various shades of orange according to the density of their population. "As soon as the heat maps began to come out, everybody recognized this is a game changer in how we look at animal populations and their movement," says Fitzpatrick. Such a big picture of what birds are doing— where, in detail, the species are going across the hemisphere, and when—simply hasn't been available before.

The eBird information is an extraordinarily powerful tool. For example, a land planner can inquire about where to build new houses to avoid threatened species of birds. Its most essential use, though, may be using birds to monitor changes in the well-being of all biodiversity, similar to what Hutto is doing, but on a vastly larger and more accurate scale. If there's a sudden drop in the number of eastern meadowlarks in upstate New York, for example, red flags go up to indicate that critical habitat might be in trouble, along with rare flowers or mammals. Biologists can then go in and see what's up.

The application isn't perfect in how it counts birds, though its accuracy is a matter of debate. The engine that makes eBird data usable is machine learning, or artificial intelligence—software and hardware that sort through all of the disparities, gaps, and flaws in data collection and figure out ways to make it better. "Machine learning says, 'I know these data are sloppy, but fortunately there's a lot of it,'" says Fitzpatrick. "It takes chunks of these data and sorts through to find patterns in the noise. These programs are learning as they go, testing and refining and getting better and better."

Sightings are collected by bird-watchers, instead of experienced researchers, so they are likely not precise. To fill that gap, top birders from Cornell travel around the world to train people like Martinka in methodology. There are also some five hundred volunteer experts who read the submissions for accuracy. About 2 percent get rejected. Rare bird sightings also get special scrutiny.

This high-tech sentinel system that tracks birds is now being used to help the birds themselves as the planet hurtles toward the next great extinction. Combined with radar of birds in flight, weather information, and even the calls of migrating birds captured by microphone arrays on the tops of buildings, eBird provides a detailed map of their whereabouts. Knowing precisely where, when, and how many birds are moving up the East Coast during their migration, for instance, helps managers know which electricity-generating windmills might need to be shut down or which lights in which office buildings should be turned off, since very large numbers of birds collide with both.

Something known as a pop-up wetland may be the most creative use of eBird. The 450-mile-long Central Valley in California used to be one of North America's most productive wildlife habitats, dotted with marshes and swamps, laced with creeks and rivers. In the last century, though, this sprawling valley has been almost entirely converted to farms, its natural habitat decimated. But eBird provides a creative way to create strategic new bird habitat here. I visited the pop-up wetlands with Mark Reynolds, who works for the Nature Conservancy and is one of the main designers of a pop-up wetland program called BirdReturns. As we drove through field after field of plowed-up farmland and precious little in the way of forest, grassland, or marsh—in which birds thrive—he explained how it works.

Come spring, hundreds of thousands of dunlins, sandpipers, snipes, whimbrels, marbled godwits, and fifteen other species of shorebirds, many of which are in decline, start their long-distance migration north from Mexico to Canada. Along their way on this arduous journey they need places to rest and feed. And one of the critical stopping-over places is the Central Valley.

The problem is the lack of good places to feed when they get there. Buying valuable agricultural land for birds would be prohibitively expensive, so the Nature Conservancy found a way to create new habitat in the cheapest way possible.

As the migrating birds reach the United States, sightings gathered on eBird from bird-watchers provide a very precise accounting of their progress north. In the days before they reach the Central Valley, sushi-rice farmers in the birds' flight paths are paid by the Nature Conservancy to flood their fallow fields. By the time the birds start dropping in to the Central Valley, a newly created wetland is waiting for them. "The water needs to be between two and four inches deep," Reynolds explained, as we pulled up to a broad, recently flooded rice field filled with thousands of birds called dunlins. Any deeper and the birds' legs aren't long enough for them to wade and forage.

I watched as these shorebirds zoomed into the wetlands and waded on stiltlike legs through a few inches of water or across glistening mudflats to ferret out worms, insects, crayfish, and snails with their long, slender, specially adapted bills. When the clouds of birds land during migration, it's as if the landscape comes alive, a kaleidoscope of sizes and colors. Beautiful as it is, it's a cold, cruel world out there for migrating birds, as they fight winds, weather, and predators, so finding good habitat in which to eat and rest along the way is essential.

Because BirdReturns pays only for several weeks of water, when it's essential, it leverages very modest amounts of money for these temporary stepping-stones rather than incur the expense of purchasing them outright. And the program is nimble, the location of the pop-up wetlands easily adapted as weather, migration routes, or commodity prices change. "It's a new Moneyball," Eric Hallstein, an economist with the Nature Conservancy, told me. He was referring to the Oakland A's and their data-driven approach to creating a winning baseball team on a relative shoestring budget. "We're disrupting the conservation industry by taking a new kind of data, crunching it differently, and contracting [for protected land] differently."

It's long been recognized that birds bring us a great deal of important information about the world—and in that regard we don't give them their due. "The ability of the birds to show us the consequences of our own actions is among their most important and least appreciated attributes," wrote the naturalist Marjory Stoneman Douglas in 1947. "Despite the free advice of the birds, we do not pay attention."

Thanks to eBird, that is changing. Now more people—a lot more—are paying attention to birds for their free advice, and it has injected new purpose into birding. "People for generations have been accumulating an enormous amount of information about where birds are and have been, jotting it down in notebooks," Fitzpatrick says. "Then it got burned when they died." No longer. The vast amount of data that birders collect because they love birds is now being put to use not only to protect birds, but to maintain the well-being of all of nature, including ourselves.

CHAPTER 4

# A Murmuring of Birds: The Extraordinary Design of the Flock

Birds are the magicians of the nature! They are here, they are there and they are everywhere!

—MEHMET MURAT ILDAN,
TURKISH NOVELIST

On a nickel-gray fall day, while I was with Mark Reynolds, the biologist with the Nature Conservancy, reporting on pop-up wetlands, I was witness to my first "dance of the dunlins." As I stood watching, the dunlins suddenly left the still, mirrored surface of a pond and went airborne, erupting as one into a great undulating cloud. Like an aerial marching band, they first created a teardrop shape, then suddenly turned and shifted in unison to create a flat block shape, and then rolled over into a circle. They changed color, too, first flashing gray feathers in unison, then brown and then white. I stood there in open-mouthed silence, mesmerized by one of the most wondrous of nature's wild bird mysteries.

Biologists believe these murmurations, as they are called, must be in some way vital to birds' survival. Why else would birds spend so much energy forming them? The most common theory is that such swarming is a defense against predators, a constantly shifting set of group eyes watching for danger as well as a flock making a much bigger and stronger target than single birds. The constant twisting and turning may also be a defense mechanism that confuses those who seek to eat the birds. Or it could be that these formations are some sort of complex signaling, a way to show other birds where to roost or feed. It might be a way for leaders to assess the condition of the flock—if, indeed, there are leaders. One researcher who specializes in the study of animal swarms believes that the dance of the dunlins, and other flocks, may both be a product of, as well as generate, what is called *metacognition,* a collective mind that is much bigger and smarter than the sum of its parts. "The group can sense the world and solve problems the way individual components cannot," says the biologist Iain Couzin, one of the world's leading swarm theorists. The director of the Max Planck Institute for Ornithology in Germany, Couzin studies bird flocks and other group animal behavior to better understand the principles underlying a broad range of other group behaviors. "For example, we ourselves are a collective of cells," he says. "How do organs form and function together when there is no overall form or blueprint?" At the heart of the mystery, he believes, is the transfer of information among members of the group.

How the mesmerizing order of a murmuration arises out of the chaos of thousands of birds has baffled and inspired humans for centuries. In recent years it has piqued the interest of physicists, aeronautical engineers, mathematicians, hedge fund managers, sociologists, computer scientists, and others who would like to understand what turns the random gathering of many thousands of birds into the exquisitely coordinated movement of a seemingly single being. "An evening murmuration is more than just the dance of starlings," write Andrew J. King and David J. T. Sumpter, two scientists who study the enigma. "It is a glimpse into one of the fundamental motions of life."

The study of murmurations is part of a larger field of study called "swarm intelligence" or "collective animal behavior"—the spontaneous synchronous movement of schools of fish, herds of mammals, swarms of bees or locusts, and other animal groups. The property of a higher order arising out of seeming randomness is a phenomenon called "emergence," and numerous scientists are trying to get to the bottom of it. "This is the big, big mystery of science," says Steven Johnson, a science journalist, media theorist, and author of a book on the subject called *Emergence*. "It's bigger than black holes and bigger than Superstring."

Cracking the secret of emergent behavior and flock intelligence could throw open a brand-new understanding about the principles that govern the world around us. Just a rudimentary understanding of the self-organization of a bird flock (as it is also sometimes called) has already proven useful to better predict bird strikes on aircraft, to provide new ideas about how traffic moves on highways, to help physicists decipher particle swarms and how crystals form, and to enable engineers to enhance the control of remotely piloted aircraft. And experts who are mining this complex natural phenomenon for ideas say that this is just the beginning. It could someday cast light on the phenomenon of embryogenesis, the way the myriad cells come together right on cue in the symphony of developing life to form the liver, heart, and other organs in a growing embryo. It might lead to the creation of medicine-carrying nanomachines that could be deployed to swarm through the veins and arteries of the human body to target a range of illnesses from muscle disorders to cancer, or it might even lead to a far more refined understanding of how our brain works. Understanding how bird flocks self-organize has led engineers to develop experimental fleets of unmanned drones and robots for search-and-rescue missions, and scientists speculate that it could one day be used to help us better understand how people make economic decisions or why they vote the way they do, or to help steer rampaging fans to safety at concerts and sporting events. And flock intelligence is the basis for a seven-thousand-person collaborative computer project aiming to harness the metacognition of the world's best and brightest in the hope of finding a supersolution to the colossal problem of climate change.

To solve these problems we need to understand the physics underlying the spectacle of many thousands of birds perfectly executing split-second and unanimous decisions. Some liken murmurations to the flow of water. Some say it's a phase transition, equivalent to the hardening of water into ice. Others say bird flocks behave like sand particles sliding downhill—one or two begin and then exponentially impact others, creating a cascade. Others compare bird flocks to iron filings that align themselves and are drawn, in symmetrical groups, to an unseen force.

One of the biggest mysteries is how the information about what the flock is going to do next is transferred from bird to bird in the tiniest fraction of a second. And how do they do it without colliding with one another? Through visual cues? By using airflow from the movement of neighboring birds? By tapping in to some sort of telepathic ability? Through quantum mechanics?

It could be any of these things or something completely different and as yet unknown. Iain Couzin has studied the swarming of locusts, as well as birds and fish. Among the insects, he found that the collective movement was driven largely by a fear of being eaten by a fellow locust. When their abdomen was de-enervated so that they couldn't detect movement behind them and react fearfully to the encroaching cannibalism, the swarming broke down. But that is only the case with locusts. Something else, something different, experts say, is going on with birds.

Getting to the bottom of emergent behavior would likely tell us something about the evolution of bird cognition. The southern weaver bird and bowerbird, for example, both build elaborate nests with sophisticated designs. Are these avian architects capable of higher order learning and memory? Or are they merely displaying rote behavior, placing sticks in a certain pattern, with the elaborate nest then arising from these repetitive movements?

One of the most compelling discoveries beginning to come out of flocking research is the nature of the information that birds transfer among themselves: how they make a decision on when to migrate, for example, or how they decide on a food source. This may not only lead to a better understanding of emergence, but perhaps advance new ideas about the ecology and conservation of animal collectives, whether flocks of songbirds or herds of antelope. "The group is informationally very, very important, especially during migration," says Couzin. It is so vital that it may mean the difference between life and death, not only for an individual, but for an entire species.

One of the first people to notice and write about the concept of emergence was Sir Francis Galton, a nineteenth-century polymath who studied everything from anthropology to statistics. Galton made his observation of collective intelligence not among birds but in a flock of British fairgoers. In 1906 some eight hundred people entered a contest to guess the weight of an ox. While no one person came close to the actual weight of 1,198 pounds, when all of the guesses were tallied and their mean value assessed, the figure was just a pound off—1,197. No other guess came close. Other experiments on collective intelligence have been repeated again and again with similar results.

It was another Brit, Edmund Selous, who became the first modern scientist to delve into the phenomenon of murmurations. Born in London in 1857, the British ornithologist and writer meticulously observed murmurations and other types of flocking for thirty years, filling notebooks with precise and poignant descriptions of the maneuvers. "Each mass of them turned, wheeled, reversed the order of their flight, changed in one shimmer from brown to gray, from dark to light, as though all of the individuals composing them had been component parts of an individual organism," he wrote.

Selous roundly rejected the notion of a flock leader, because the birds moved so synchronously. He concluded there were only two possible causes for murmurations: the appearance of a predator, or something he called "thought transference"—an unseen telepathic connection between birds. While birds often rose in unison in response to a predator, he observed, they also often rose for no apparent reason, as if "interconnected by some rigid arrangement of invisible wires." Only some type of telepathy could account for such high-speed movement, he wrote, "thought transference so rapid as to amount to practically simultaneous collective thinking." His book on the subject is titled *Thought-Transference (or What?) in Birds*.

Another early theorist was Wayne Potts, a biologist at Utah State University. High-speed films of dunlins, the long-legged shorebird I saw in California, were slowed and examined carefully. Potts, too, concluded there was no leader. A sudden shift in the direction of the flock could start anywhere, he said, and then ripple away from its point of origin and through the rest of the flock almost instantaneously. A big question in Potts's hypothesis was how the wave moved from bird to bird at an average speed of 15 milliseconds—far too fast for a bird to react to the movement of a bird next to it. Tested in a laboratory, dunlins needed 38 milliseconds to respond to a flash of light. Yet within the flock they somehow reacted in less than half that time.

Chorus lines provided Potts with an explanation. Experiments in the 1950s showed that unexpected dance moves passed from dancer to dancer at a rate of 107 milliseconds, nearly twice as fast as the human visual reaction time of 194 milliseconds. The explanation for both bird and dancer alike, Potts surmised, in a paper titled "The Chorus Line Hypothesis," is that they must anticipate the arrival of the spreading wave and react before it gets there.

The advent of computers brought a new opportunity to solve the mystery of the flock. In 1987 a software designer, computer programmer, and animator named Craig Reynolds came up with a novel approach. Reynolds worked at a now defunct computer technology company in Southern California. "There was a cemetery next door and they had huge flocks of blackbirds and I would go there and watch them," Reynolds told me. The real birds that swooped, dived, and soared above the headstones inspired him to create a flock of animated birds. Reynolds called his virtual creatures "Boids" and programmed each of them using agent-based modeling, a standard method of computer modeling that sets complex behavior in motion using only simple rules. It means that each bird or "agent" gets only a few basic rules, with no higher-level programming, and that in turn dictates the emergent group behavior. Reynolds came up with three simple rules for Boids: *collision avoidance,* meaning that the individual birds must always stay separate; *velocity matching,* which means that the birds must fly at the same speed; and *cohesion,* which means they steer in the same direction as their flockmates and move toward the average position of the flockmates around them.

Others have built upon Reynolds's pioneering work, and the models have gotten more lifelike as computing power has increased. "Scientists measure their worth in terms of citation and the citations of that Boids paper are in quite a few fields," Reynolds told me. "From collective robotics, to crowd modeling, to evacuation scenarios in architecture, there are quite a few things that trace their roots back to that original paper."

The approach is widely used by the motion picture industry, and it shows up in the design of the computer-generated swarms of bats and armies of penguins marching and flying through the streets of Gotham City in Tim Burton's 1992 *Batman Returns.* Other recent manifestations are the dramatic battle scenes in the *Lord of the Rings* and *Hobbit* film trilogies, which used something called MASSIVE software—Multiple Agent Simulation System in Virtual Environment. Each orc, elf, and other animated creature was an independent agent programmed with a set of rules similar to Reynolds's Boids, though twenty years later they were far more sophisticated, and each had the ability to see, hear, and move. Each agent guided itself only by those principles—the animator wasn't involved.

That is why the battle scenes in those movies took on a remarkably lifelike appearance—because each warrior was imbued with an artificial life of his own. Reynolds won an Academy Award in 1998, in part for his bird-flock-based computer designs and their contribution to film animation.

Finding the key to the riddle of flocking has been elusive because we lack the ability to recreate the behavior in three dimensions, in part because moving from two to three dimensions is exponentially more difficult. "I created real flocking for nonreal creatures" is how Reynolds explained his two-dimensional approach to me.

A team of European physicists, economists, and biologists tried to remedy the 3-D problem with a well-funded project that ran from 2004 to 2007 called STARFLAG—Starlings in Flight. Cameras that took pictures of the birds from two different angles to create a three-dimensional view were set up on the roof of the Museo Nazionale Romano to study some three thousand urban starlings swarming in the skies over Rome's Termini railroad station.

I talked via Skype with Andrea Cavagna in Rome, a statistical physicist who studies "disordered systems"—how elements interact in ways that are unpredictable—and who led the project. The first discovery Cavagna and the team made is that while the murmurations are large, they are deceptively flat—like a potato chip, he said, even though they look like a potato. The brass ring when the project began was to find and describe the fundamental principle of emergence in birds—which would be a new force in physics—and then see if it applied to such things as financial markets and stadium riots. The STARFLAG team did not reach their goal, and Cavagna has doubts about how widely the principle of emergence in birds can be applied elsewhere.

"Collective behavior in most systems I study has been shaped by evolution," he says. "Whether it's birds, fish, sheep, or midges, what you see are the products of a very long process of evolutionary tuning. When you look at collective behavior in social systems, you don't have the backup of evolution. Looking at how people get out of a stadium in a panic—a hundred thousand trying to get out a door in ten seconds—you can hardly argue that this behavior has been shaped by evolution. It's collective behavior, but very different than a flock. Same with finance, how can you argue that [financial trading] has been evolutionarily shaped? The mechanisms are probably very different."

STARFLAG was not a complete flop, though. The biggest lesson learned from it is that distance between birds is not the key factor in how birds carry out their murmuring, as was surmised, but rather the topology, or spatial relation, of the birds in the flock. That is, each bird is positioning itself based on six or seven other nearby birds, no matter their distance. This was a new principle in the world of physics. "The birds don't care at all about the distance of their neighbors, only the number of neighbors," Cavagna said.

After three years of study the team came to a familiar conclusion: Birds are far smarter than we have assumed. "An interaction based upon the number of neighbors, rather than their distance, implies rather complex cognitive capabilities in birds," said Irene Giardina, a STARFLAG researcher.

Understanding the dynamics of flock behavior and how birds use it can greatly enhance efforts to protect the birds. The decline of the group mind may be behind one of the world's great extinctions.

A single humongous flock of passenger pigeons that flew over Ontario, Canada, in 1866 may have been one of the largest flocks of birds ever witnessed—it's estimated to have contained more than three billion pigeons. The dark bird cloud was a mile wide and more than three hundred miles long, and it took fourteen hours to pass a single point in southern Ontario. The din of flapping wings would drown out other sounds when such a large flock flew over. The artist John James Audubon described one flock he saw in 1813. "The air was literally filled with pigeons," he wrote. "The light of noon day was obscured as by an eclipse, the dung fell in spots, not unlike melting flakes of snow."

The massive flocks of passenger pigeons that once lived throughout North America were the largest assemblies of any animal on the planet, second only to locusts. Now the bird is extinct, and Iain Couzin thinks that one of the things that may have been its undoing is a loss of the great diversity of information that a flock contains, and a diminution of the emergent properties.

Though Couzin works in Germany, he is Scottish by birth, and I reached him via Skype in Scotland, where he was working as a visiting professor at the University of Edinburgh. The role that information plays in a flock—or a herd or a swarm or school—is likely essential to the survival of the individual as well as the collective, the soft-spoken, bespectacled scientist told me in his brogue. "The informational components really haven't been considered until now," he says, "and we are just starting to get the technology and theories to get data from migratory systems."

When the number of any animal that travels in a group falls, he says, there is a loss of information, because each individual carries its own unique knowledge based on experience. Knowledge about survival is a precious commodity because bird flocks are noisy places, informationally speaking. The critical signals a bird needs to receive to make its way in a perilous world can be drowned out by a lot of other distracting and unimportant factors, or "noise"—both natural and man-made. Birds navigate in migration by processing a multiplicity of cues, from stellar constellations to the smell of a forest or rain to a sense of the earth's magnetic lines. Wind, thunderstorms, artificial lighting, and a range of invasive odors, for example, can easily drown out the guiding signals. That's where the sound or sight of another migrating bird, or many of them, before or during flight, adds strength to other signals. "They have to decide when to go, when to take off, and when to rest," Couzin tells me. "If you are doing this by yourself, and if you are trying to pay attention to climatological information and other cues at the same time, it's potentially very, very noisy. So what groups of animals are doing when they gather to prepare to migrate is casting their ballot," Couzin says. "Birds and fishes effectively vote; it's a group vote and they are amazingly efficient voters. They can make consensus decisions."

By "voting" he means that the groups gather and use bird communication to decide which wheat field to feed in that day or when and where to migrate. Then inexperienced birds follow their elders until they become experienced and can cast their own vote. Birds use social cues to vote, which may be a glance from another bird, the posture of their body, how they hold their head, how far away they are, or the arrangement of the network of fellow birds.

"A group can have a sense of environment that is far more developed than the individual sense of the environment," Couzin says, which is the very definition of emergence—akin to the way the group of fairgoers could guess the weight of the ox, but no individual did. A flock of birds is far better equipped for the rigors of migration than individuals. Whether it's obvious, or subtle and unconscious, communication within a group is essential for survival.

Adding to the complexity of these group dynamics is the fact that these flocks and swarms are nonlinear—that is, a very small increase in the number of birds can make the emergent properties far more robust, or a small decline in their numbers can dramatically reduce the strength of the signals critical to the successful completion of their perilous mission.

What may have sealed the fate of the passenger pigeon, then, Couzin says, is a breakdown in emergent properties—the metacognition—of the entire flock. It may have diminished their group survival skills and doomed them long before Martha, the last passenger pigeon, keeled over in her cage at a zoo in Cincinnati in 1914. "Perhaps they simply didn't have the capacity for normal behavior, whether reproduction or feeding—at lower critical density," hypothesizes Couzin. As the number of birds dwindled, so, in effect, did their perceived options.

The fate of many species may depend on how effectively the group generates emergent information. What effect does the fragmentation of habitat and reduction in the flock size of warblers, for example, have on the survival of the flock during the perilous time of migration, and perhaps, if their numbers are low enough, the species? At what point does the flock lose enough of its smartest voters to no longer be able to make the right decisions about where and when to migrate? We are a long way from knowing. But Couzin has some speculations.

What might a murmuration of starlings be communicating, verbally and nonverbally, as they morph from shape to shape in the skies above Rome or the Scottish Highlands? "They produce these large flocks typically in the autumn, when their foraging is biased toward agricultural lands," Couzin says, and over the Skype screen he stops to underscore the fact that this is only his informed speculation. "Finding the food source is quite difficult, but when you get there you've got a whole field of stuff to feed on. In that type of environment, the body mass of these birds can change by a third within a day. They can really lose energy if they seek and fail to find food."

Most of the time, he says, a bird is surrounded by fellow travelers who are in a similar condition. "If those around you have done badly, you don't want to pay attention to their social cues," Couzin says. "You want to know how the rest of the flock is doing, the ones not nearby, what condition are they in and what do they have to say about food sources." Birds get this vital information likely through a combination of social cues, whether the smell on another bird's breath, the direction from which they arrived, or their overall physical condition.

The murmuration itself, then, "a drunken fingerprint rolling across the sky" as the poet Richard Wilbur put it, is "a stirring of the flock, it's like stirring your coffee," says Couzin. The real information exchange occurs the following morning, "when the birds wake up near a new neighbor." So the murmuration "allows them to make better information as social groups." It's best not thought of as a sharing of information, though, or as an all-for-one-and-one-for-all spirit. "It could be stealing information as well," he says. "There's no such thing as the benefit of the group. Or benefit of the species. It's individuals maximizing their fitness, for their own benefit." In other words, there's no altruism here, just each bird looking out for itself, struggling for survival in an often perilous world.

Thomas Malone is one person who hopes that the metawisdom of bird flocks can also be coaxed out of a human collective. He runs the Center for Collective Intelligence at MIT, and their core research question is how people and computers might be connected to spark a human metamind. "Perhaps birds can help us answer that question," he says. "By looking at the way birds work together, connected only by their own sight, sound, and perhaps a sense of touch, how can that teach us about new possibilities for connecting people?"

Google and Wikipedia are examples of a metacognition that flows from individuals working separately, Malone says, and the entire Internet, in a sense, is a giant flock of humans contributing toward a whole. "The Google algorithm harvests and analyzes that knowledge in a way that creates new knowledge that didn't exist before," ranking Web pages by what is most important, in the case of Google, or in the world of Wikipedia creating an entire online encyclopedia with many people entering small bits of information. Creating this metacognition, he says, has great potential.

A project under way in Malone's center, called Climate CoLab, links seven thousand people—world climate experts, professors, grad students—in a large network. The participants offer up solutions for different aspects of a problem, on which the entire group then votes. "Climate change is a complicated, hard problem," says Malone. "But we know we have a way to solve complicated, hard problems, and that is with collective intelligence."

Swarm intelligence might also shed light on unrecognized dimensions of our democracy. In one of Couzin's study of swarms, in this case fish, his team came to the surprising conclusion that the uninformed members of a collective may actually perform a great service. "A strongly opinionated minority can dictate group choice, but the presence of uninformed individuals spontaneously inhibits this process, returning control to the numerical majority," says Couzin. If two groups in a bird flock are at loggerheads about which direction to go in, then the few individuals who know little and have no strong preferences can listen objectively to both sides and sway the argument toward the best one. "The uninformed act as a social glue," Couzin says.


There's a scene in the science fiction film *Minority Report* in which Tom Cruise's character, John Anderton, is in harried pursuit by police. They think they have him cornered in an abandoned building, so the cops release tiny spiderlike robots into the house. The heat-seeking swarm believes he is submerged in a bathtub of ice water, and when a single bubble of air escapes from his nose, they sense it, realize where he is, and shock the water until he emerges. This concept is not just science fiction—robotic swarms are under active development, a design based on the intelligence of real swarms.

While these robotic swarms are not yet smart enough to catch criminals, they might soon be able to stop pollution. "Say you wanted to clean up oil spilled in the Gulf of Mexico with underwater robots," says Cavagna, who has done work on robotics. "I can control all of them centrally from Houston, but that's a lot of work. Or I can give some distributed rules of interaction," akin to the Boids model of simple commands. "All of you stay in an overall radius in the Gulf of Mexico, and stay a certain distance from your neighbors, never too close, never too far, so the system self-organizes so there is an even spread of units. The whole system rebalances and all interact together."

Imagine that you are in charge of that couple of dozen bots scouring the ocean floor. Suddenly one dies—an "external perturbation," in swarm lingo, akin to a falcon taking out your starling neighbor. Instead of Houston having a problem it has to deal with from a

distance—remotely rearranging the surviving robots to accommodate the gap—the robots have been programmed to behave like a
real flock of birds and rearrange themselves automatically. This way
"the perturbation is local and the reaction is local and the problem
doesn't distribute automatically to the rest of the flock," Cavagna
says. "The controller doesn't need to take action. A swarm that has
distributed control is much more robust than a swarm with central-
ized control."

Not everyone believes that the answer to the why and how of bird
flocking lies in the field of physics or in simple commands that create
complex behavior. Could it be that Selous was right and the glue
that keeps a flock together resides along the lines of mental telepathy? Rupert Sheldrake, a Cambridge-educated biochemist who often
ventures into explanations that are outside the realm of mainstream
science, thinks so. His signature idea is something called *morphic
fields*. The concept has been roundly attacked as scientific heresy by
some—including Couzin, who calls Sheldrake a charlatan—while
other scientists believe it casts innovative and important light on
how the world may be governed by unseen fields of intelligence.
Does a pea plant grow because there is something entirely within the
DNA that governs its growth, or is there an intelligence that guides
the plant's growth through the double helix of the DNA, which
serves as a kind of antenna for this information? Sheldrake believes
it's the latter, and that is what he refers to as a morphic field. "They
underlie the organization of proteins, cells, crystals, plants, animals'
brains and minds," he writes. These fields occur simultaneously on
different levels. One governs the bodily form of an organism, for
example, and another, the social field that coordinates "the behavior
of individuals within social groups," such as flocking birds.

In his book on animal telepathy, *Dogs That Know When Their
Owners Are Coming Home,* Sheldrake writes that if the birds were
simply responding visually, it would "require the bird to be able to
sense, notice and react to such waves almost immediately, even if
they are coming from directly behind them. This would require them

to have practically continuous, unblinking three-hundred-and-sixty-degree visual attention." His morphic field hypothesis "would make it easier to understand how the birds not only perceive and respond to the maneuver wave as a gestalt, a combination of form and wholeness. Through it they could grasp the movement of the flock as a whole and respond to it in accordance to their position within it."

Quantum mechanics may well be where the answer to this perplexing puzzle lies. If a bird's eye is entangled with the earth's magnetic lines in a quantum way, as one leading theory on migration holds, the minds of the birds could also appear to be separate yet somehow working together. Or there is another possibility. In post-STARFLAG work, Cavagna found that when a flock of four hundred birds turns, there are no laggards at all; it happens very precisely in less than half a second. And he found that if there is a misalignment in the flock, and a few birds don't correct themselves, that misalignment has to travel throughout the whole flock. Cavagna's team worked out equations for this behavior and found that it matches precisely the equations for the movement of quantum superfluids, in which all particles align and behave the same way.

Couzin, though, thinks it is something far simpler. A recent study by his team tracking the visual fields and body postures of fish shows that some members of the group are "socially influential," as well as susceptible to social influence and aware of outside influences such as predators. The rest of the group is watching and responding to them. These key players don't travel in the rear of the group, which is where predators focus their efforts. "We expect this applies to birds," he said, "but we do not yet know if that is the case."

We are clearly a long way from fully understanding murmurations. And despite the public fascination with their wild murmurations, starlings are in steep decline across Europe. In 1979 the Royal Society for the Protection of Birds began a citizen science program that asked Britons around the country to methodically count birds in their yards. The average number of starlings observed when the program began was fifteen per yard; it is now three. Observers have also been counting fewer birds in the murmurations, and so the bird is now on the red list of concern. Dunlin populations are also dwindling. Will the diminishing numbers of these birds cause the loss of the flocks' emergent properties? "If murmurations do indeed function as collective information processing, it's distinctly possible," says Couzin. To protect social species in peril, he says, it is essential to understand not just numbers and habitat, but the critical role of information and the metamind. "Modification of the environment by humans is unlike anything else that has occurred in the evolution of these migratory birds," says Couzin. "So we are having a direct impact on their social behavior and the way they interact with the planet." And then climate change is coming down on top of these degraded ecosystems. "If we push them too far, like we did with the passenger pigeon, we can find ourselves reaching one of these non-linearities where a tiny bit more influence may cause a very sudden collapse."

We have much to learn from birds, and the discovery of an unknown force of nature is surely one of the most important. That's why protecting ecosystems is far more than just protecting a bird. It is harboring the almost unimaginable potential to learn more about this fascinating world we inhabit.

# PART II

## The Gifts of Birds

# CHAPTER 5

⌇

# The Power of a Feather

Hope is the thing with feathers that perches in the soul—
and sings the tunes without the words—and never stops
at all.

—EMILY DICKINSON

Birds provision us—that is, they provide us with things we need. One of the things they have long provided us with is their feather cloak. The long arc of evolution has created one of nature's most exquisite and versatile designs, a natural technology that elegantly and efficiently accomplishes a wide range of vital functions that humans have not come close to replicating.

All birds, and only birds, have feathers, and in each type of bird, their feathers have uniquely evolved for different needs in different ecological niches. The feather needs of a flightless ice-floe-dwelling penguin, for example, are very different from those of a peacock that lives in the tropical forests of southern India.

Researchers are only beginning to understand how extraordinarily complex and sophisticated a technology the seemingly simple bird feather really is. I've just been reading about feathers, and as I sit here at my desk holding a peacock feather that is nearly weightless and a foot long. I realize that I have never really come close to seeing—not seeing, *perceiving*—what a feather really is. I have, of course, held many, worn them in down jackets in the cold northern climate in which I live, fly-fished with them, and admired their beauty, but as I hold this feather now and stroke its silky, zipperlike barbs I realize I have never truly appreciated the wonder of a feather. And as I will soon learn, I don't know the half of it.

Feathers, like hair, are dead. And like hair, they are made of keratin. It is one of nature's toughest proteins, both extraordinarily sheer and light and yet incredibly strong, essential to giving birds their incredible flying agility—feathers are even more integral than muscles to sustained flight. Despite their near weightlessness, they are one of the most effective insulators known to humans; they are used in the warmest clothes we make. And they are durable enough to protect ground birds as they flee predators through dense, lacerating tangles of brush and grass. Their ebullient colors—the most vibrant colors in the natural world—are the ultimate eye candy, critical to both sexual attraction and effective camouflage, and so they make great imitation insects for fishing as well as extravagant clothing for everyone from exotic dancers to tribal kings. And they are renewable—each year, birds get a fresh set, and if some are damaged, they do a repair molt, shedding them to make way for new growth.

Richard Prum is an expert in the evolution of the feather, and like many others in the field, he is a self-described bird nut. "There are bird nuts that grow up into scientists," the Yale ornithologist says. "That's deeply a part of who I am. The work I do is just bird nuttery gone mad." His passion for feathers, he says, is part of this mania, and he talks excitedly and rapidly about them, spicing his words with numerous expletives.

"I was teaching ornithology," he recalls, about the moment when the miracle of feathers dawned on him. "I put feathers on the syllabus and thought I understood them. As I prepared for lectures on how feathers grow and evolve, though, I read the literature and it absolutely blew my mind. 'This,' I thought, 'is really cool.' And I had this messianic desire to share it with my students." And as fascinating as feathers are, says Prum, there is still much about them that we don't know.

But how complex can this simple, gracefully arcing peacock feather be? As I look it over and ponder, it seems pretty straightforward. Where does it hide its secrets?

The long feathers of a soaring eagle are very different from the small feathers of the flightless penguin that needs them solely for warmth. Still, there are just two main types—vaned feathers, the long version that cover the body's exterior, and down feathers, which cover the body beneath the vaned feathers. Some birds have another type of feather, called semiplumes, which are rare; they are as fine as hair and grow at the base of the down feathers for warmth.

The large vaned flight feathers that support thrust and lift when the bird is in the air are called *remiges* and are fastened firmly to ligaments, which in turn are fastened to the bones. *Primaries* are connected to the *manus,* the "hand" of the bird, on the outer part of the wing, while *secondaries* are affixed to the bird's ulna, or forearm, on the inner half of the wing. Tail feathers, or *retrices,* are used for both control and stability during flight. On the edges of both the remiges and retrices are rows of smaller feathers called *coverts,* which streamline the wings and insulate.

There is one other, much rarer type of feather. Parrots, herons, and a few other species of birds have something called "powder down" feathers that produce a fine, oily dust that sifts through the feathers, a self-cleaning mechanism that gathers up moisture and dirt, which the birds can then preen away. Birds also have glands that produce preen waxes, which coat feathers and allow them to better shed water and keep themselves dry.

Each feather is attached to a follicle, the so-called "goose bump" that birds have, which are arranged in neat rows on the skin. Most feathers have a long stiff shaft with two vanes that extend out from it. The central shaft is hollow at the end—called the *calamus*—where there are no vanes, and solid where the vanes are attached to the shaft, which is called the *rachis*. The vanes are asymmetrical. The leading edge of the feather in flight, the outer vane, is the shorter side, while the longer, trailing side is the inner vane.

Individual barbs, the long strands that extend from the shaft and make up the vanes, are similar to tree branches that grow progressively smaller as they grow farther away from the trunk. Each barb has a barbule on the tip, which is hooked, and on the barbules there are microscopic barbicels, also hooked. This hooked design in the exterior feathers creates a gently fixed, self-assembling and disassembling wing, a kind of natural Velcro. As a bird flies, the feathers are constantly fastening themselves together and coming apart, firmly hooking up on the downstroke for better resistance to air, to create lift for the bird's flight, and unhooking to yield to resistance on the upstroke, making the transition effortless and efficient.

One known but virtually uninvestigated secret of bird feathers is how they function as a living part of the bird, acting like transducers that sense wind speed, pressure, and location and relay this real-time flight information to the bird's brain, via a few specialized nerve cells at the point where the quill attaches to the skin. There are tiny, adroit muscles around each follicle that allow feathers to be microadjusted based on flight conditions.

Another secret is how, precisely, wing feathers keep birds in the air. "Equations for how air flows over feathers are based on the idea that a feather is solid, like an aluminum airplane wing," says Prum. "But in fact the feathers are made of tiny microscopic air gaps and huge volumes of air inside the feather. All of our analysis about bird flight ignores the fact that the feathers are porous. Why? Because no one has measured it."

And feathers are so strange in the way they grow that they are a prime example, along with such things as eyeballs and flowers, of how novelty arises during the course of evolution. A decade ago, most researchers believed that feathers, in the dinosaur past, were reptile scales that had somehow turned ragged. "Feathers grew like shingles on a house and suddenly caught air to give the bird gliding power and ultimately became branched and feathery," says Prum. "It's a crock. It couldn't happen that way. All you had to do was know even fundamental things you could teach somebody in five minutes about how they grow and they'd say Wow, that theory can't work."

It can't work because as a feather emerges from the skin, it begins its new life as a tube, and eventually unfurls to become a feather. "When it unfolds, the inside of the tube becomes the undersurface of the body of the feather, and the outside of the tube becomes the top surface of the feather," says Prum. "It's simple geometry—you can't get to a feather from a scale because a feather is a tube." Feathers on dinosaurs—and some think all dinosaurs had them—first functioned as something other than an appendage for flying, such as for warmth or protection.

Feathers are weird in other ways. "Feathers are branched like a tree," says Prum. "But they grow from their base like a hair. Unlike a tree or a plant, which branches at the tips so the trunks give rise to branches, which also give rise to twigs, feathers grow in the opposite fashion. The tips of the feather are older than the nodes; in other words, the fusion of the trunk to the branches is the last thing that happens. Imagine a tree with all the twigs and tips coming out of the ground and the trunk is only formed at the end of the process. How the hell does that happen? How the hell could that happen!"

In feathers, then, is a lesson: "about how we reconstruct a path across all of evolutionary biology—relevant to eyeballs, relevant to flowers, relevant to limbs, to hair. All of these innovations are novel,"

Prum insists. That is, they arise on their own, not from something else. "With evolution, shit happens you could never imagine."

Wherever or however it originated, the feather is a miracle, so light it allows birds to soar effortlessly through the air. And the substantial size of golden eagle feathers allows them to soar above all other birds. That's why many North American native people say the golden eagle bridges two realms, heavenly and earthly. It flies higher and therefore it sees and hunts better than any other bird, and so it is master of the world below; and at the same time it flies so high that it is considered by many to be a spirit bird. "The feather is sacred because the golden eagle flies high up there and takes our prayers farther up into the sky," says Lee Plenty Wolf, a spiritual teacher in the Oglala Lakota tribe. "The eagle is our messenger to Creator."

The fantastically light natural technology of the feather allows the bird to hunt with unparalleled ferocity. A twelve- or fourteen-pound eagle can even kill a small deer. With only an occasional flap of its wings, the massive bird glides low, perhaps ten or fifteen feet above the ground, just over the herd. When the eagle finds the most vulnerable deer, often one isolated from the others, it lands atop the animal and pushes its powerful two-inch talons into the deer's withers, which is the high point between the shoulders. As the bleating fawn runs wildly, the eagle maintains its grip, like a rodeo rider in a life-or-death match. They sometimes even fly off with the animal, then drop it to its death.

These traits of the golden eagle are why its feathers were—and still are—the greatest of spiritual medicines and the most sacred of objects for Native Americans. For centuries, young men have waited in mountaintop eagle catch sites, holding a rabbit as they crouch in hiding, sometimes for days, then grabbing the eagle when it lands to take the bait. Tribes used the sacred feathers in long trails on war bonnets, to decorate ceremonial objects such as peace pipes and staffs, and to wrap into medicine bundles, collections of sacred objects. They wore feathers in their hair, in armbands, on their legs. They placed them on altars in sweat lodge ceremonies or at sun dance rituals that are at the heart of indigenous spiritual beliefs. They sought—and still seek—through their use of feathers to leave the world of earthly concerns and connect with the power of the golden eagle, the spirit realm, and the Creator. Many young Lakotas are making their way back to their traditions, Plenty Wolf says, "away from alcohol and the gangster life, and when they go on vision quests they earn plumes from the bird. The feathers guide them to be part of the higher realm in order to speak and pray to Creator."

Feathers from an eagle, when presented to someone, represent the highest values of trust, bravery, and honor. "A long time ago, if a young man won a horse race or went on a war party or did some other great deed, he was given an eagle feather," he continues. "Nowadays if a person graduates high school or goes into the military and is in combat, they get one. It's how we honor what they have done."

Others prize feathers for more down-to-earth reasons. Among some of the possibilities that are being explored by researchers and biomimeticists—people who seek to learn the secrets of nature to design better products—are the following developments:

- Penguins are covered in a thick mat of feathers attached to muscles that the waddling bird tenses and locks down, creating a covering so warm it allows them to plunge into icy water hundreds of feet below the surface. These feather mats could help design lightweight panels for watercraft and waterproof textiles.
- Most birds have minuscule grooves around their feathers that trap a blanket of air beneath them in order to prevent moisture

from binding to the surface, an adaptation that has engendered ideas about waterproof and self-cleaning fabrics—a "smart" umbrella that perfectly sheds water and stays dry.

- Some pigments in bird plumage may be antibacterial to keep feather colors from degrading, which has potential for medical applications.

- The color blue in the feathers of bluebirds and peacocks and in the "flying jewels" look of the hummingbird is made by an unusual natural technology—two-dimensional crystal cells that form the way beer foam does and reflect a prismlike light back to the viewer. The technique is being used to create a car paint that seems to glow, and some researchers think a similar process could be used to create a new type of self-assembling laser.

- Researchers are trying to understand how bird feathers detect disturbances in airflow to allow the bird to alter its flight. Experimental systems have been developed for airplanes based on this natural technology to reduce in-flight turbulence.

Feathers are also the softest thing nature produces, and so duck and goose down is crammed into pillows, duvets, and mattresses. Down feathers have also long been used to make the warmest and lightest types of clothing and bedding that we humans use, whether comforters, sleeping bags, or down jackets that fend off the most frigid weather in the world's harshest environments. I didn't fully realize how down-dependent I was in my life in Montana until I wrote this chapter and did an inventory. I own two toasty goose-down jackets for the below-zero temperatures we usually get in winter, a down vest, two down sleeping bags, a down comforter, down pillows, and down bootees.

Some 80 percent of the down and feathers used for clothing and bedding is sourced these days from China, and it's a by-product of ducks and geese that are raised and slaughtered for food. Not so long ago, commercial down became controversial because it was savagely plucked from live animals, but fortunately industry standards have evolved. The clothing company Patagonia is an industry leader in this area and recently instituted something called "traceable down," a policy mandating that all down must be able to be traced back to a supplier who meets humane standards, including the requirement that feathers not be live-plucked or gathered from geese who are force-fed for foie gras.

The world's best-quality down is eiderdown, which comes from a wild sea duck, the common eider, which lives in the Arctic Circle. The drake is a gorgeous half-white, half-black bird, and the hen a less dazzling, yet still gorgeous, mottled brown and black. The big birds are well equipped with an impenetrable coat of feathers—the tough outer covering—and down—the softer and warmer lining underneath the feathers—so they can live in the harshness of the wind-whipped northern seas. After laying her eggs in a depression in the ground, the eider hen plucks the down from her breast so she can press her bare skin against her eggs to warm them, while the clumps of fluffy gray down she plucked are used to line her nest against the punishing Nordic climate.

To protect themselves from the wiles of the depredating Arctic fox, the birds nest in large, densely packed colonies on the coasts of Europe, North America, and Siberia, in some places as many as thirteen thousand nests per hectare. In an ancient and symbiotic relationship, Icelandic farmers have commercialized the eider's down. They fend off foxes and other predators while the birds nest, and then gather down through the incubation period, taking just enough so that they don't affect the hatching. Then, after the young have fledged, they gather the rest and replace it with soft grass. They don't get very much—the annual global harvest of eiderdown is between two and three thousand kilograms, enough to fit into one truck.

As nature's best insulator, easily compressed and resilient, down is a precious commodity and has not yet been replicated by the hand of man. It is rated by fill power—the higher the loft of the down, the higher its fill power. Most outdoor gear uses a fill power between 400 and 900, the most common being goosedown, which is 500 or so. That means that one ounce of this goosedown takes up 500 cubic inches. Eiderdown has a fill power of 1,200, which means one ounce fills more than twice as much volume, and traps more air. That makes it a better down, because it retains its loft longer and is both far warmer and far lighter. The rarity of real eiderdown, its superb quality and labor-intensive gathering, means it sells at a steep price—an eiderdown-stuffed comforter starts at about three thousand dollars and runs as high as twenty thousand. There is nothing quite like one, though, and they have been highly prized as far back as the ninth century, by the Vikings, who made eiderdown and silk comforters. The most delicate and lightest of fill, wrapped in silk, was all that was needed for a comfortable night, even in the most bitter Arctic winters.

Common eiders, unfortunately, are not as common as they used to be. The state of the eiders was the subject of a captivating 2013 documentary called *People of a Feather,* which examined the relationship between the Inuit people of northern Canada and the bird, whose down has been the essential ecotechnology that has allowed these people to survive there for centuries. Come winter, the birds thrive in open water in the ice, diving for sea urchins and mollusks. Warmer winters and freshwater discharges from hydroelectric dams, however, have upended the ecosystem, and much of the eider's food has disappeared. Their numbers have plummeted.

Many people are devoted to the feathers for their decorative aspects as much as they are for their warmth. On a quiet street between downtown Los Angeles and Hollywood, in the basement of a nondescript building, an unusual business has a simple sign out front with its name: MOTHER PLUCKER. As I walk inside, my eyes are bombarded with floor-to-ceiling racks of feathers dyed every color of the rainbow, hanging in long boas or giant clumps, stashed just about everywhere. Specks of feathers float in the air. I am greeted by Willy Zelowitz, who founded this establishment in the 1970s, and Lelan Berner, who has long helped him run it. Both are hippie artist types turned entrepreneurs. Willy, who sports a Hawaiian shirt, is stout and has a graying beard, and his hair is gathered in a ponytail. Lelan, with silver hair, is more outgoing, and both are ecstatic to have a chance to talk about the things they can do with feathers. They introduce me to Betty Lo, who is busily attaching white feathers to a frame with a hot glue gun. "She's the best angel wing maker on the planet," says Willy, in his soft voice. She very well may be—Mother Plucker Feather Company is one of the top purveyors in the world of feather costumes for the movie and show business industries.

Bird feathers have adorned fashionistas since the beginning of human time, made into everything from boas to plumage-bedecked hats, dresses, and capes. Native Hawaiians captured millions of tropical birds and plucked their brilliant red, green, yellow, and black feathers for their famous and striking clothing, everything from capes and helmets to leis, skirts, and sacred art. In the eighteenth century, some eighty thousand tiny mamos, or hoohoos— nectar-feeding birds with long, slender curved beaks, used like straws to sip nectar from flowers—were killed, their yellow rump feathers plucked to make a single golden-yellow cloak for King Kamehameha.

Today, Mother Plucker sells feathered costumes to a wide array of customers: showgirls in Las Vegas, circuses and carnivals, performers on cruise ships, exotic and burlesque dancers, and people who simply want to go all out for a big event. "They come in here from all over for Halloween, Burning Man, and Lightning in a Bottle," says Berner, referring, in the latter two, to popular art and alternative lifestyle festivals for which people gather in flamboyant costumes. "When Jenny McCarthy was Playmate of the Year in 1993, I did her angel wings," Willy says proudly. "And we made angel wings for Victoria's Secret models." Strippers, too? I ask. He nods emphatically. "Huge part of our business," he says.

Zelowitz shows me a long, fluffy string of maroon feathers and explains that it is a marabou, from the marabou stork. Some 75 percent of the feathers Mother Plucker uses—which, like down, are also by-products of birds raised for meat—are from turkeys, the rest from chickens and occasionally ostriches, whose feathers are far more expensive. The feathers are almost always from the males. "The boys are always prettier, the boys are beautiful," says Lelan with a sigh.

They made sixty pairs of angel wings—at $650 a pair—for Jim Carrey and a stageful of dancers at the Academy Awards. They once made a feather headdress for an elephant, not to mention the giant chicken a guy rode recently in a Jack-in-the-Box commercial; they made Nubian feather fans for a Michael Jackson video, and dozens of pairs of wings for the angels who roam a Bible theme park called the Holy Land Experience in Orlando, Florida. "It's our imagination and their pocketbook," says Willy.

Founded in the 1970s, today Mother Plucker employs thirteen workers and a row of industrial sewing machines. Feathers, it seems, never go out of fashion—for good reason, says Lelan: "It's natural. And there's nothing else that gives you the loft, the shape, the movement. They are drop-dead beautiful and nothing comes close to a real feather. You can't make it out of synthetics."

Willy nods in agreement. "Humans have always been attracted to birds," he adds. "They control the sky. That's at the heart of the attraction to feathers."

As I get ready to leave Mother Plucker, Lelan asks if I would like to try on a pair of their finest angel wings. The first pair I try, a garnet color, is too small, but the second, a brilliant and showy white, much to my surprise, fits quite nicely. I wonder what friends would think if I showed up to a dinner party in giant white angel wings.

While fashion feathers now come from domestic birds, they were once sourced in the wild, and the lust for them drove many birds to near extinction. In the 1870s an enormous global craving for feathers to adorn women's hats began and burgeoned for three decades. At their peak, feathers were worth as much as gold, twenty dollars an ounce. Many of the most valued species were found in the brackish swamps of the Florida Everglades, ground zero for plume poaching. Poachers killed herons, spoonbills, and ibis, but the most sought-after birds were snowy egrets with their brilliant white plumes, especially the *aigrettes,* long delicate trailing nuptial plumes that grow off the back of their head during the mating season.

Gathering these plumes was a gruesome slaughter, akin to the killing of the American buffalo for their tongues. The egret is an elegant, long-legged bird, about two feet tall, with a wingspan of three feet. It has a long neck and brilliant yellow feet that it uses to stir the mud to release shrimp, minnows, and fish. When poachers came upon a rookery of thousands of beautiful white-plumed egrets, the land looked as if it were snow-covered. Adult birds were quickly killed with clubs, their feathers plucked, and the carcasses left to rot. Orphaned eggs and baby birds were sentenced to death by neglect.

Unlike the passenger pigeon, which disappeared from the planet altogether, egrets were saved, primarily through the efforts of Edward Avery McIlhenny, who owned the McIlhenny Tabasco sauce company. In the 1890s, McIlhenny rounded up a handful of surviving egrets and placed them in a large "flying cage" in the middle of his 250-acre private refuge called Bird City on Avery Island near the Louisiana coast. After they acclimated and bred, he set them free to migrate, and they did, returning each year in greater numbers, until there were many thousands of birds.

But such large-scale destruction of birds is far from a thing of the past. The treatment of birds in Hawaii is the most egregious example. The precolonial demand for birds and feathers by natives was so great on the islands that it rendered many species extinct, including the gorgeous mamo, a black bird with yellow markings and a long tail and long curved beak. More recent habitat loss, along with disease-carrying mosquitoes and invasive species such as mongoose, which eat bird eggs, have accelerated the decline. A third of the endangered birds in the United States are in Hawaii, more than in any other state. "The islands are a borderline ecological disaster," says Cornell Lab of Ornithology director John Fitzpatrick.

But as we see in the example of Bird City, it's possible to walk nature back from the brink. "One of the most important messages in bird conservation," notes Fitzpatrick, "is when we fix the things we broke, a lot of things come back very quickly. There's a resiliency that's really remarkable." The miracle of a feather is something almost all of us take for granted, whether it's our warm and comfy down jacket or the angel wings we wear onstage. Understanding the prized and wide-ranging role that feathers play—and could play— shows why it's important to fix the things we broke to keep birds from disappearing.

# CHAPTER 6

From Egg to Table

## PART ONE: THE CHICKEN

We can see a thousand miracles around us every day.
What is more supernatural than an egg yolk turning into
a chicken?

—S. PARKES CADMAN, CHRISTIAN MINISTER
AND NEWSPAPER COLUMNIST

The ancestral chicken still walks the earth.

The chicken from which all chickens are descended is the red jungle fowl, and the species still lives a life untamed, prowling the green jungles of India, China, and Indonesia. This wild version of the rooster is a noble-looking bird that very much resembles its barnyard kin, with a rainbow of brilliant feathers that range from deep gold to fiery red to rich turquoise, a dangling red wattle, and the inverted Mohawk of the bird world, the red comb. "An untamable leopard," wrote William Beebe, an early-twentieth-century naturalist, as he watched a wild rooster. "Low hung tail, slightly bent legs, head low, always intent, listening, watching, almost never motionless. Just for a moment he was agleam, the sun reflecting metallic red, green and purple from his plumage." In contrast, wild females are a drab brown, but their feathers are sleeker than those of their domestic cousins.

The free-ranging life of the red jungle fowl is similar in some ways to that of birds on a farm—small, stable flocks, softly clucking as they scratch through the leaves on the forest floor, scouring the ground for seeds, insects, and fruit. They build ground nests of leaves and branches but occasionally roost in trees. They are famously shy and sensitive birds, very hard to glimpse in the wild. If they are captured, the shock they experience has been known to induce heart failure.

Cut to my hometown Costco. In the far back of the cavernous store, pale, naked, headless birds spin, impaled on giant stainless steel rotisseries, beneath the yellow-orange glare of heat lamps. These are what the red jungle fowl have become. I have been a big fan of these savory birds whirling on their culinary Ferris wheel, as have many people. Costco is one of the country's leading purveyors of cooked chicken, selling a whopping sixty million birds a year. The bird even has its own Facebook page and consistently makes Costco's top-ten bestseller list. A delicious three-pound whole cooked chicken for five bucks—what's not to like?

When someone raises the subject of birds, the term "food source" might not immediately come to mind. Birds, however, are one of the most consumed food sources in the world. Where would the world's pantry be without the chicken and its eggs? Globally more than 40 billion chickens are consumed each year, and the United States leads the flock of chicken eaters with more than 8 billion eaten annually. Every day in this country 217 million eggs are fried, scrambled, poached, basted, souffléd, soft-boiled, hard-boiled, poured into batter, made into salads, and otherwise devoured. Chickens are an ideal food in many ways: cheap, plentiful, nutritious, and rib-sticking. That's why breeding them has become a massive industry, and that is why there are far more chickens in the world than any other bird. In North America alone there are 10 billion chickens, compared to 4 billion wild birds.

I have always eaten chicken, and I love chicken cooked just about every which way. But as I researched this book, my feelings about the role of these once wild birds in our lives has come up for a serious reexamination. In the chicken I see a parable of what we humans have done to the world—poisoned our food, and ourselves, to keep things cheap; denied the true costs of our actions; demeaned, com-modified, and tortured animals despite growing evidence that they may have a mind not radically different from our own. In the chicken, too, is a tale of how we have lost a sense of what's sacred.

The chicken was first domesticated in India in 3200 B.C., though there is evidence that the modern-day chicken has multiple genetic origins. In the days before refrigeration, chickens were a mobile, self-replicating source of meat and eggs, and they were a common com-panion of pioneers, explorers, and the military. From its jungle home in Asia, the bird spread around the world on sailing ships; the chicken may have come to North America with Columbus on the explorer's second voyage, or it could have been brought by Polynesians who reached South America's Pacific shores.

Since they came in from the wild, chickens have been an essential food source in almost every culture. As far back as four thousand years ago, Egyptians built mud-brick ovens connected by small tun-nels to a massive system of incubators, filled with ten or fifteen thou-sand eggs, which assured a steady supply. The eggs were kept at the requisite temperature of 99 to 105 degrees with carefully tended straw and camel dung fires; their minders would test the temperature of each egg by pressing the shell gently to their eye, the most sensi-tive part of the body. The Romans—who are credited with the invention of the omelet—also built elaborate chicken structures, often above bakeries, and used the heat of the bread ovens to warm the eggs and chicks. Wood smoke was also piped in to fumigate the chickens and kill mites and other parasites.

Chickens were, and still are, honored in many cultures. Hens have long been a symbol of fertility and roosters a symbol of virility. Like many other birds, they've been revered as a mediator between the mundane and the sacred. Across Asia, Europe, and the Middle East, the rooster was a solar symbol because it crowed to greet the light of the rising sun, which banished evil spirits. In Japan, the sacred white rooster still has free run of Shinto temples because it is thought to wake the sun goddess with its call. Chickens have been viewed as prognosticators, most notably in Rome, where chicken augury was a serious business. Military expeditions would pack chickens along with them on their journeys, to be watched over by augurs who carefully interpreted their feeding behaviors to divine such things as whether the gods favored an upcoming military engagement or political action. Just before launching the battle of Drepana, Sicily, in 249 B.C., against the Carthaginians, Publius Claudius Pulcher, the senior Roman commander, consulted his chickens. They refused to eat, which was a bad omen that frightened the soldiers and crew. Ignoring the sign, Pulcher dumped the sacred chickens overboard into the Mediterranean, saying, "Let them drink if they don't wish to eat." The Romans were crushed in the battle.

In Santeria, an African religion combined with elements of Roman Catholicism, orishas are spirits that represent God, and they sometimes require a blood sacrifice. To this day it's often done with chickens. The ritual slaughter is carried out with great respect for the animal, adherents say, by carefully severing the carotid artery so that the bird passes out before it is slaughtered. As the blood spills onto the ground, the orishas are said to feed on it. Some Jews also have a form of ritual chicken slaying, on Kapparot, the eve of Yom Kippur, the holiday of repentance. Observant Jews recite prayers as they wave a chicken over their head, believing that the bird takes on their sins and the chicken bearer is granted atonement. Afterward, the bird is slaughtered.

Eggs, too, have been long and widely regarded with spiritual reverence, a powerful symbol of the mystery of creation. The World Egg is an archetypal motif for the beginning of the universe, found in cultures across the world, from the Finns to the Chinese to the Greeks. Numerous origin stories feature a dark and forbidding universe traversed by a cosmic bird that drops its egg into the chaos and brings light. Egyptian myth holds that the earth came into being from the waters as a mound of dirt, atop which a mythical bird laid its egg. Inside the egg was Ra, the sun god. Jews eat eggs at Passover as a symbol of mourning, but also as a symbol of the cycle of birth and death. Christians celebrate the resurrection of their savior with dyed eggs because they are a symbol of fertility and rebirth.

Yet today, birds are not regarded as lofty or mythical. Instead, over the last few decades, the wild jungle fowl has been genetically engineered through breeding to be little more than a meat-and-egg-producing machine that can, by cutting numerous ethical corners, be mass-produced cheaply.

Until the 1940s, chickens were raised primarily for their eggs. Chickens themselves assumed a starring role in the American diet during World War II because chicken, unlike beef and pork, wasn't rationed. Consumption has soared since, from about forty pounds per person a year in the 1970s to nearly ninety today. The market is worth some $60 billion, most of it consolidated in the hands of a few giant "integrators"—massive corporations that control with great precision all phases of the chicken's life, from birth to table, a process that guarantees low prices and maximum profit. The chicken, as a result, is the most industrialized animal in the history of the world.

Most of the American chicken industry is concentrated in the "Broiler Belt," which extends from East Texas throughout the Southeast and up the East Coast to the Delmarva Peninsula, where Delaware, Maryland, and Virginia come together. The modern-day eating chicken, called a broiler, was invented in 1923 under the aegis of one Mrs. Wilmer Steele, of Oceanview, Maryland, the first person to raise large numbers of chickens for their meat. The wife of a coast-guardsman stationed at Betheny Beach, Maryland, she was raising a few chickens when five hundred chicks were accidentally shipped to her, instead of the fifty she had ordered. She built a bigger coop, raised the chickens, and made so much profit on the two-pound birds that she doubled down. By 1926 she was able to build a broiler house with room to keep ten thousand birds. That First Broiler House still stands at the University of Delaware, enshrined as a kind of monument to the mass production of chickens.

A hurdle for the burgeoning industry was the fact that keeping too many chickens together indoors caused disease. The invention of antibiotics in the 1940s changed the equation, and, along with synthesized nutrients such as vitamin D and folic acid, allowed mass production to scale up. And lo and behold, it was discovered that, for some reason, antibiotics also help chickens pack on the pounds faster, almost miraculously so.

Perdue, Pilgrim's, Tyson, and other industry giants that are re-ferred to as Big Chicken are why the bird can still be bought for less than a dollar a pound, and a fully cooked three-pound chicken at Costco is priced at only five dollars. It's not just efficiency of scale. The rock-bottom cost of chicken is made possible by corporate sleight of hand combined with a ruthless efficiency. The real costs of raising that chicken are hidden to the consumer. Producing these cheap birds is done at the expense of humane treatment of the crea-ture, the economic subjugation of the families who grow them, and the low-wage workers with high rates of injuries who process them.

Many farmers who contract with the big integrators complain that they are manipulated and controlled through a de facto chicken monopoly stacked entirely in the favor of the integrators. The small-scale chicken farmers must assume large debts to build chicken houses and get into the business, and then their payment is deter-mined solely by the integrator company. In an industry that controls everything from egg prices to the retail product, there is little in the way of a free market to set prices. "Almost invariably, from everything I've seen, the farmer loses," said Christopher Leonard, a journalist who wrote an exposé of the Tyson company in a book called *The Meat Racket*. The system "keeps farmers in indebted servitude, living like modern-day sharecroppers on the ragged edge of bankruptcy." In fact, more than two-thirds of the people who grow chickens live below the poverty line.

Manure from chicken house waste in the Broiler Belt is another of the industry's egregious problems. Growers in Delmarva shovel nearly two billion tons of manure out of their chicken houses each year. Some growers spread the powerful fertilizer across farm fields, but the amount of poop is so massive and expensive to move that most of it is left to pile up. "Legacy manure" has accumulated for decades, polluting the region's waterways with pulses of phosphorus and nitrogen with each rainstorm. The fertilizer causes the runaway growth of algae in waterways, which consumes most of the oxygen, creating "dead zones" where fish, shellfish, and other aquatic life simply cannot survive. Chicken manure is one of the largest contributors to the ecological nightmare that the Chesapeake Bay has become, where abundant populations of blue crabs, oysters, and striped bass in one of the richest and most productive estuaries in North America have been decimated. The growers don't have the resources to clean it up, and the industry doesn't consider it a priority.

What Big Chicken has meant to the quality of life of the descendants of the noble, free-ranging jungle fowl may be the biggest controversy of all. Modern-day broilers are a white-feathered chicken, usually Cornish Rock crosses, selected over the last half century to put on weight at blinding speed and optimize the amount of grain they eat, something known as a "feed conversion ratio." A chicken needs a bit less than two pounds of feed to put on a pound of meat, whereas it takes seven pounds of feed to create a pound of beef.

In the modern American chicken house, broilers are raised in cavernous five-hundred-foot-long metal buildings crammed to the rafters with chickens. A farmer who raises the birds for companies such as Pilgrim's or Perdue typically houses twenty-five thousand chickens in a shed and has around twenty sheds. And the term "squalor" doesn't begin to describe life inside a broiler house. They are cramped, hot, and sunless, and the filthy birds can barely move—they can only eat, which is the way growers want it. As a result, the birds reach their full weight of up to five pounds in just five weeks, as opposed to the natural chicken, which takes six months to grow to full weight. I toured chicken houses in Georgia, and the air was filled with choking clouds of dust and fetid with ammonia fumes from feces and urine. Many chickens have sores and injuries; mortality rates are so high that dead birds litter the floors and are thrown into heaps.

Because weight gain is most important and white meat is prized, genetics have been modified to produce grotesquely oversize breasts, which severely strains many birds' developing legs and pelvises, causing them to become deformed and buckle beneath their own weight. Millions of chickens die from coronary failure, because their hearts are too small to pump enough blood into their excess flesh. The journal *Poultry Science* calculates that if humans grew at the same rate, they'd weigh 650 pounds by the time they were two months old. That weight is put on by the cheapest feed possible.

For many the system is indefensible. "This must constitute in both magnitude and severity, the single most severe, systematic example of man's inhumanity to another sentient animal," wrote John Webster, a professor emeritus in animal husbandry at the University of Bristol in the UK.

The industry not only causes suffering in chickens. A recent study by *Consumer Reports* found that a whopping third of grocery store chicken breasts are contaminated with salmonella, a bacterial disease that hospitalizes some nineteen thousand people a year with cramps, diarrhea, and, in some cases, far more serious problems. Salmonella kills more people than any other food-related illness. Experts say the chicken industry is poorly regulated; accountability in the vast labyrinth of the federal food safety system is low, and so not only is food poisoning a problem, but when it does occur, weak laws and oversight mean integrators often avoid accountability, recalls, or punitive measures. Big Chicken is also breeding superbugs with the routine use of subclinical levels of antibiotics. The widespread use of the antibiotic Cipro in chicken feed, for example, creates antibiotic-resistant bacteria, which are passed on to consumers. This bacteria causes extreme illness and even death, and also leaves some people immune to conventional antibiotic treatment.

The half million or so American workers who process the birds are also victims of Big Chicken. Working on automated processing lines where 145 dead birds fly by per minute, the workers suffer injuries at twice the national average of other laborers. The most serious wounds are deep cuts, amputations, and musculoskeletal disorders from the hundred thousand repetitive motions a worker makes in an eight-hour shift, all for some $25,000 a year. Unions and other organizations have petitioned the U.S. Department of Agriculture for new safety regulations to stem the problem; Big Chicken, for its part, has instead asked the USDA to allow them to increase the number of birds on the line to 175 per minute.

The high speed of the line also exacerbates cruelty. Live birds have their legs shackled and are hung upside down to travel into an electric stunner that kills them; after that, an automated blade decapitates them, and from there they are placed in hot water to scald them and remove their feathers. The speed causes poor placement of the birds, and results in some 825,000 chickens a year that miss the stunning and decapitation and instead are scalded alive, a far more painful way to die. "You can't stop the abuse at these speeds," Mohan Raj, a British poultry slaughter expert, told The Washington Post. "It's so fast you blink and the bird has moved away from you."

Egg laying, too, left its pastoral setting behind long ago. Each year more than ninety billion eggs are laid in the United States by more than three hundred million laying hens. While the red jungle fowl lays ten to fifteen eggs per year in two clutches, the hens of the egg industry are bred to average an annual output of 271 eggs. The birds that do the laying are stacked in columns of wire-floored battery cages. Crammed into a tight space the size of a piece of paper, the bird's natural habits of nesting, dust-bathing, and perching can't be performed. Her beak, the primary tool with which a hen pecks about and otherwise engages the world, is severed to keep her from injuring herself and neighboring birds. Since male chicks are useless as egg layers, they're killed after they hatch, placed on a conveyor belt and sent to be gassed or minced in a grinding machine—a process sometimes called Instantaneous Mechanical Destruction, or IMD, which, poultry farmers argue, is humane.

All of this continues despite growing evidence that chickens are no dumb clucks. They're no raven or crow, but some have a complex language with at least two dozen separate and remarkably specific terms to describe different situations—there's a different alarm call for a threat coming by land or coming over water, for example. Chicks have a distress call—loud chirping or peeps with descending frequencies—that means "lost" or "cold" or "frightened." There's a fear trill and a pleasure trill, while the hen calls her chicks with soft repetitive notes. A "tidbitting" call tells the chicks where the mama hen is feeding and which foods to eat.

Chickens can solve problems and display an ability to think about the future. Birds in lab tests who are placed in front of a button that offers either a few grains right away or more food if they wait longer pick the button for delayed and greater gratification. They have good memories. When it comes to maintaining their pecking order, hens can remember more than a hundred faces and the rank of each bird. "Perhaps most persuasive is the chicken's intriguing ability to understand that an object, when taken away and hidden, nevertheless continues to exist," says Dr. Chris Evans, a professor of psychology who has studied chickens at Macquarie University in Australia. "This is beyond the capacity of small children."

The idea that birds and other animals are conscious beings with some semblance of emotional lives that deserve greater ethical consideration is gaining support these days. Research on the mind of birds is taking us straight toward a collision with a very large question: At what point are we no longer morally justified in genetically warping them, keeping them in filthy mass-production facilities, and grinding them up to be turned into Chicken McNuggets?

Karen Davis moved to a small house in Montgomery County, Maryland, in the 1980s, seeking a rural way of life. Her landlady was raising a hundred or so broiler chickens crammed into a nearby wooden shack. A recent transplant to the country, Davis was drawn to the soft clucking of the birds and started walking down the leafy path from her home each day to watch the broilers. "I noticed how fast they grew, and how large, and how crippled they were," she says. One day she walked to the shed and discovered that her newfound friends had disappeared. But there was a lone shadow visible through the window and she peered inside. "There was a little chicken who had been left behind," she told me. "It was a hen, and she was totally crippled and stained with dirt and feces."

Davis fell in love with the little bird, whom she brought home, cleaned up, and named Viva. The hen inspired in her a vow to liberate the industrial chicken. She founded a group called United Poultry Concerns, based at her farm in Machipongo, Virginia, at the southern end of the Delmarva Peninsula, in the midst of the teeming chicken slums at the heart of the broiler industry. She maintains a small refuge where, at this writing, ninety-four rescued broilers and other chickens, and a lone guinea hen, are living out their natural lives. "Chickens represent the largest number of land animals being raised and slaughtered for a human food source," she says. "Because they are exempted from anticruelty laws, I felt that chickens needed a voice and organization focused on them and their plight."

And because of people like Davis, some things are changing in the chicken and egg industries. As of 2012 the European Union has banned the use of battery cages for egg layers, in favor of "enhanced cages" large enough for a hen to spread its wings without touching another hen or the sides of the cage. California also voted to end the use of battery cages, and because no eggs can be sold in California that are not from the more humane cages, the rest of the country will likely have to follow suit. Federal legislation is currently being considered.

Broiler reform has been more elusive. The highly profitable integrators have been resistant to change because it would require a massive overhaul in how they do business. Instead they have become more creative with labeling, calling birds "natural" or "humanely raised" when they are not. But there is a rapidly growing public awareness of food issues and ever stronger demand for sustainable, healthier alternatives, and many consumers no longer accept business as usual, for their own sake and for the sake of the chicken.

The first inklings of real change seem to be under way. Tyson Foods, the largest U.S. poultry producer, recently asked its growers to adopt the five freedoms of animal welfare—freedom from hunger and thirst, freedom from discomfort, freedom from pain and injury, freedom from fear and distress, and freedom to express normal behavior. In 2016, Perdue Foods announced a major overhaul. It would put windows in its growers' dark and dingy chicken houses, give the birds room to move, and put them to sleep before they slaughtered them. United Egg Products, which represents 95 percent of the egg production in the United States, has vowed to end the practice of destroying male chicks by 2020 by creating females only in the egg. It remains to be seen how thoroughly the industry carries out these reforms.

It might seem odd, but there is also a conservation issue regarding chickens, even with forty billion of them in the world. For thousands of years peoples in valleys and villages around the world selected chickens for the tastiest meat and best egg production, coloration, and temperament, for how well the chickens were suited to the local climate, and for their resistance to local diseases. This human genetic selection created local breeds, which are also known as land-

races, such as the White-faced Black Spanish, Javas from the Far
East, the Sussex and Dark Cornish from England, and the Chinese
Cochins. Now many of these local breeds have disappeared, dis-
placed by the uniformity of Big Chicken. Just as there are endangered species in the wild, there is endangered biodiversity in the
barnyard. It's potentially bad news for the future of a bird that because of our dependence on it just might be what some call "the
most important bird in the world."

The birds of Big Chicken "are compromised," says Jeannette Beranger, who works for the Livestock Conservancy, which searches
out and preserves rare poultry breeds. "Their energy goes into growing big and growing fast," while the genes for other traits, such as
heat tolerance or disease resistance, have shrunk vastly, placing the
birds at risk of epidemic or extinction in a changing, warmer world.
Researchers are trying to head off such problems. A team from the
University of Delaware recently traveled to Africa to search out a
breed of chicken called the African Naked Neck to see if its ability
to withstand hot temperatures could somehow be bred into broilers
and help the chickens raised by the industry to survive as the climate
warms.

Experts liken the loss of these genes for survival to the destruction of a library without knowing what's in it. That's why the Livestock Conservancy searches out genetic diversity in chickens, ducks,
turkeys, and geese before these local breeds blink out, and it encourages farmers to raise those birds. "There's no way to recreate their
breeds genetically once they are gone," Beranger says. "You never
know what you are going to lose until it's too late." Twenty-nine
poultry breeds are currently ranked by the conservancy as critical,
which means fewer than five hundred individuals of each breed remain.

It's no small irony that the most serious threat to the ur-chicken,
the red jungle fowl, is from Big Chicken. Genetically pure wild red
jungle fowl are hard to find. Common chickens are wiping out the
genetics of their wild ancestors in South Asia with their domesti-

cated DNA, in which some of the wild attributes have been lost. And red jungle fowl DNA might one day prove crucial to salvage the genetically inferior domestic chicken. In fact, in the 1950s a researcher went to Asia and brought back a flock of purebred red jungle fowl that he released near Fitzgerald, Georgia. The Georgia flock still wanders the town's streets and parks, and may help provide the needed genetics to help restore the wild jungle fowl.

Recently, as I wandered through my local Costco, I stopped to look at the rotisserie chickens they were selling. A flock of cooked birds swam in their warm, fragrant juices in the hard plastic shell in which they are sold. A pang of sadness washed over me. We have taken an elegant, free-ranging wild animal, sacred to much of the world, bred it senseless, and turned it into a grotesque abomination. God knows where these chickens had been or what had flowed through their systems. Were the chickens I was eating causing me to become more resistant to antibiotics? Did they carry salmonella? Had a slurry of their waste flowed into the Chesapeake Bay and helped destroy it? Were low-wage workers injured during their processing on a grim and cruel assembly line? I now knew too much. Big Chicken has sucked the soul out of the wild red jungle fowl, and in that moment, I knew that I had bought my last Big Chicken. Going forward, I'd find a source for an occasional free-range chicken—more expensive, to be sure, but a bird that had been accorded some dignity and lived a good life.

# CHAPTER 7

~

# From Egg to Table

## PART TWO: WILD BIRDS

Tame birds sing of freedom. Wild birds fly.

—JOHN LENNON

Staring into the Ping-Pong-ball-size eyes of a bird that is just a few feet away and as tall as I am is an unsettling experience, especially if it's a bird that could, if it wished, easily kill me. Emus are ostrichlike birds from Australia, gawky, pencil-necked, and flightless, with incredibly long and strong legs that can rip out a fence—or a person's jugular. They do not attack people very often, but this one, with its wide, oil-black eyes and spiky dark "hair," looks straight at me, unblinking. Is he just curious? Or is he sizing me up the way a velociraptor back in the Jurassic period might have? Meanwhile, a mob of

drumming females circles me warily. Drumming is an eerie noise female emus make with an inflatable sack in their throat, which sounds
like banging on the bottom of a plastic pail. I am told it's a courtship
sound. Are they in heat? I hope not. Fortunately, I am able to sneak
out of the pen unharmed, but when I go eyeball to eyeball with an
emu I see, more clearly than I have in any other bird, the dinosaur
lineage. It's a far different experience from when I cradled the tiny
hummingbird in my hand.

Emus are classified as ratites, a group of mostly large, flightless
birds that includes ostriches. They once had a much larger relative,
the ten-foot-tall elephant bird of Madagascar, which laid an egg the
size of a football, but it went extinct in the seventeenth or eighteenth
century, likely at the hand of humans. While ratites don't fly, they do
run. Emus gallop at speeds up to thirty miles an hour, zooming
across the dusty outback like giant roadrunners, swallowing plants,
seeds, insects, and once in a while tilting back their head to send a
lizard sliding down their long throat. The emu is one of Australia's
wildlife icons, and along with the kangaroo appears on the country's coat of arms. It is also at the heart of the aborigine creation
story, in which the emu goddess-bird drops one of her giant black-
green eggs in the air, and in a scintillating, fiery explosion it creates
the sun and illuminates the glories of Earth. The bird is memorial-
ized by a constellation called Emu-in-the-Sky.

Australian aborigines have hunted the flightless emu with spear
and noose since the dawn of time. The role of the emu in their lives
was similar to that of the buffalo in the lives of the American Plains
Indian—their quarry provided everything from food to tools to spiritual sustenance. Native hunters cut off the lean, dark meat to roast
and eat, and they stripped the hide to be made into clothing. The
feathers became earrings and other decorations, and the bones were
made into tools and jewelry. The yolks of the giant smooth black
eggs were eaten, the empty shells used to carry water. Large chunks
of the fat that nourishes the bird during lean times were saved by the
hunter, melted over a fire, and smeared on his body. After wearing

the bush medicine for three days the hunter would wade into a river to wash it off, and the cuts and abrasions he had accumulated from rugged desert life were healed.

While the great majority of meat from birds worldwide is sourced from chickens, many other birds play a nutritional role. The emu provides one possible model for how we might raise birds for food in a more humane and sustainable way.

To find out more, I drove to the tiny, rural western Montana town of Hamilton, an hour or so south of Missoula. On the edge of town was a small sign that announced the Wild Rose Emu Ranch, a typical suburban ranch-style home with twenty acres around it. Joe Quinn, a retired school principal, tall, with thinning white hair, answered the door and asked me in. He explained that he and his wife, Clover, raise a hundred or so of the big birds there.

Emus are bred commercially for their lean meat, their chicks, which are sold to other emu ranchers, their feathers, their soft, supple hide, and especially the oil refined from their fat, which is used to heal burns and abrasions, eczema and psoriasis, as well as to relieve fever, coughs, and pain. Eggshells are sold to collectors and artists.

While Joe helps out, Clover, a short, reserved woman who speaks slowly and softly, runs the operation. She ushered me out the back door of her house and took me on a tour of the ranch. The birds are kept in large pens with six-foot-tall woven wire fences. There is plenty of room for them to roam, mate, and raise their babies in the shadow of the saw-toothed Bitterroot Mountains. As we neared the pens, I watched the gangly birds run along a fence, surprisingly fast and agile. As we drew closer, the birds grew nervous. Despite the fact that the birds are a foot taller than she is, Clover is at home as an emu wrangler—most of the time. As the young male emus come into sexual maturity, they get rambunctious. "They get really spicy," Clover explained, rolling her eyes. "They want to show how mature they are, so they peck on anybody who is vulnerable and they run at you and bully you, really. They want everyone in the pen to know they are Mr. Macho."

We went into the small building where the incubator is kept. "Emus lay their eggs here in the late fall," she said, about one grapefruit-size lunker every three days. In the wild, an emu dad gathers eight or ten eggs into a clutch in a nest made of grass, sticks, or sand and sits on them for seven or eight weeks without eating. Here, Joe and Clover gather the eggs and set them into a climate-controlled incubator, where a mechanical rocker gently rolls them, just as the birds do in the wild. Seven or eight weeks later the nascent chicks start to peck away bits of shell to begin an arduous journey. Struggling toward daylight, the wet and frail chick must complete its birth process within four days, or it will run out of the vital nutrients contained in a sac in the shell and starve to death. Clover spends much of her time during the hatching season in spring bending over the eggs and coaxing the chicks by mimicking the peeps and chirps of a mama emu.

Not all eggs hatch. Some just don't develop, others are taken by predators. Here at the Wild Rose Emu Ranch they occasionally end up on a breakfast plate. Joe Quinn whipped up a whopper of an omelet for us out of just half an emu egg (one emu egg equals a dozen chicken eggs). Emu egg shells are too thick to crack on the edge of a mixing bowl, so he drills a hole with a small hobbyist's drill and then sprays compressed air into the shell, which forces the yolk and albumen out of the shell in a long string and into a small bowl. He scrambles the egg to a slight froth with a fork, mixes it with a little water, salt, and pepper, and tops it with cheese. The omelet is a little denser than one made with chicken eggs, and delicious.

Emu eggshells are one of nature's jewels, with three layers of color—black-green with specks of turquoise on the outer shell, teal in the middle layer, and then white. When the eggs are carefully emptied, there are artists around the world who carve them, creating beautiful and complex three-dimensional works of art. Clover is one of these artists, and her egg carvings were once displayed at the White House. "We got to meet Laura Bush," she said proudly.

The Quinns were part of the great emu boom of the 1990s. Emus were then thought to be the next big thing, a sure way to get rich selling birds, eggs, and meat, the Quinns explain to me over the kitchen table, but it turned out to be a pyramid scheme. When they were first imported to the United States in the early 1990s, a breeding pair of emus sold for $45,000. By 1996 the number of emu farms had exploded and a breeding pair sold for just $1,000. "The following year they were giving them away," Clover says. The emu economy had collapsed. There were once 5,500 emu operations in the United States. The best guess now is that there are between one and two thousand.

The meat of an emu, very lean, dark, and savory, something like duck meat, is marketed by health food stores as being heart-healthy, but it hasn't caught on in a big way. Because emus don't fly and their pectoral muscles—unlike the breasts of chickens—never developed mass to power their flying, there's only about thirty pounds of meat on each bird. What saved the industry from total collapse is emu oil, which sells for about ten dollars an ounce. One bird produces enough fat to make about 250 ounces. The oil is widely used for eczema and rashes, and in shampoos and ointments. For scrapes, abrasions, and burns, Clover says, "it is better than Mother's kisses."

As I researched the human relationship with wild and semiwild birds, I found it remarkable how many people still rely on wild birds for food. Today, many of us pick up our birds in the grocery store in antiseptic plastic wrap and Styrofoam, a type of gathering that is light-years removed from the people who daily go out and gather wild birds, or their eggs, or their nests. The Inuit of Alaska hunt ducks and geese; the Ju/'hoansi people of Botswana use foot snares to capture guinea fowl; the Chukchi, native people of Greenland, lower themselves on frayed ropes over the side of precipitous wind-whipped cliffs to collect prized black-and-turquoise guillemot eggs; the Kung bushmen of the Kalahari have a relationship with ostriches similar to that of Australia's aborigines with emus. In the Arctic, native Greenlanders make a peculiar raw bird dish called *kiviak,* stuffing several hundred of the seabirds called auks into the hollowed-out body of a dead seal, stopping it up tight with seal fat, and then letting the delicacy ferment under a pile of rocks. A year and a half later, when it is good and ripe, they dig it out. The well-aged bird is so soft that every part of it can be eaten, save the feathers.

A traditional bird harvest takes place off the coast of Scotland, where each August a group of men from a cluster of small villages called Ness set sail for a large rock called Sula Sgeir. With long poles that sport a noose at the end, ten hunters spend weeks clambering over the steep cliffs to pluck from their nests some two thousand young, plump gannets, referred to by the locals as *guga.* After capture, the birds are killed with a swift rap on the head, plucked, salted, and taken back to the Isle of Lewis, where they are prized as a delicacy. Despite the fact that the bird looks like an oily rag, the taste is said to be delicious—half salted mackerel, half duck—and they are shipped to gannet-loving expats around the world. Although they were once gathered throughout the UK, the Sula Sgeir harvest is the last stand of this ancient tradition.

Wild birds provide us with some of the most unusual foods known to humankind. Some Asians, for example, crave a delicacy—called *balut* in the Philippines or *hot vit lon* in Vietnam—of an almost fully developed duck fetus, in the shell, tossed back with a little salt and pepper and lemon. The ortolan is a songbird that was long served in France, captured live and then drowned, marinated in brandy, roasted, and presented to diners who ate it, bones and all, with the bird's head shrouded by a napkin. It is illegal to serve ortolans now, but top French chefs are lobbying for the birds to be allowed on the menu again.

The most precious food birds produce, though, is not their meat but their nests. There are thirty or so species of swift, and they are common in the Philippines, Thailand, Vietnam, Indonesia, Burma, Malaysia, and Singapore. One of these is the swiftlet, the cousin of the chimney swift. About the size of sparrows, though with longer tails, swiftlets have short bills and long, slender bodies, so slender they are sometimes called "flying cigars." They are, as their name suggests, wickedly fast and dexterous, reaching speeds up to a hundred miles per hour while they chase and dive-bomb insects. Like bats, they use echolocation to hunt. And they are the providers of a rare and peculiar avian delicacy called bird's nest soup. Loaded with potassium, calcium, and other minerals, the soup is thought to be a holistic tonic that is good for the skin, throat, and lungs; it supposedly boosts the libido and is purported to have antiaging and anticancer properties as well. Demand has soared, and it's now being added to candies, coffee, and even skin creams.

Swiftlets live in limestone caves along the coast of southern Asia, particularly the South China Sea. They nest in colonies and build their graceful and translucent cup-shaped nests high on the cave walls in order to thwart predators. Their building material is the spider-silk-like strands of their own saliva, and it takes about thirty-five days to create one nest. Swiftlet nests come in three colors: white, red-orange, and black. The reddish nests, the most prized, are found only in Thailand. Once thought to be colored red by the blood of the builder, the nests actually take on the color of iron, which leaches into the strands of hardened saliva from the minerals in the cave wall. A kilo of white nest sells to suppliers for $2,500, while red nests fetch $10,000.

Swiftlet nest soup is referred to as "the caviar of the East" and is quite expensive; a single bowl can fetch as much as a hundred dollars. Most swift nests are shipped to the United States and Hong Kong, and the Hong Kong Chinese are the largest consumers of the gelatinous delicacy, a hundred tons per year. The price has led to counterfeit swift nests, and the Malaysian government now uses tiny radio transmitters to track legitimate shipments.

Part of the reason for the high cost is the difficulty of gathering nests that lie more than two hundred feet above ground. By torchlight, traditional swiftlet nest harvesters ascend the dark, damp cave walls on rickety, primitive scaffolding made of bamboo and vines. One-man harvesting can gather fifty or sixty nests a day. There is growing concern about the ecological impacts of the harvest; one study found that the swiftlet population declined by three-quarters in Southeast Asia as a result of overharvesting and habitat destruction. Today, soup nests are often farmed along the sea by entrepreneurs who build concrete structures to create a safe place for the birds to build their nests.

Most of the modern world has lost the everyday connection we all once had with birds. In the 1990s I spent several weeks among the Yanomami tribe of Brazil, considered the last great aggregation of Stone Age peoples in the world, and observed firsthand how differently humans relate to birds when they live without walls and windows and cities to separate them. Birds are everywhere, always. They surround the *yano*, the doughnut-shaped communal living structure, and during the dawn chorus they fill the jungle with cacophonous calls, screeches, and singing. For the Yanomami, birds are sentinels, serving as a way to track game or a noisy alarm to warn of the approach of danger.

Several Yanomami I met kept wild toucans and parrots as pets, their wings clipped and their legs tethered to a stick, and wild bird eggs are vital as food for the tribe. One fellow I hunted with killed a couple of chicken-size birds I had never seen before, which he lured by mimicking their call. Birds are especially prized among these people for their beauty and the grace of their feathers and wings. The Yanomami adorn themselves daily with a rainbow of vibrant bird feathers, from scarlet and emerald parrot or macaw feather fans to the brilliant yellow feathers attached to the slender stick that they pierce through their nose. The hunter I befriended killed a gorgeous turquoise and black hummingbird and promptly stuck two clusters of its feathers in pierced holes in his earlobes. Some Yanomami wear whole tiny birds in their ears. During the several days of celebration

I witnessed, they slathered sticky black tree sap on their heads and stuck fluffy white bird down to it, creating a kind of party hat. In their culture, as in many cultures, birds represent a connection between heaven and earth, and wearing them is an attempt to ascend to a higher realm.

I have long hunted wild birds, mostly pheasants, sharp-tails, and occasionally waterfowl, though I do not, if at all possible, rise in the predawn blackness in cold and snarling winter weather, as one needs to do to properly hunt them. There is a singular reward from shooting wild birds on the wing, and it is different from other kinds of hunting, more akin to the aesthetic experience of fly-fishing. My primary quarry is the common pheasant, or ring-necked, which is derived from the green pheasant, a native of China and so an avian interloper here in the United States, though now well established thanks to the passion of hunters. It is not a subspecies but what's called a hybrid swarm, the result of many different genetic elements of captive breeding. In Asia, where the bird is revered as a symbol of cosmic harmony, there are thirty different subspecies, including the golden, cheer, green, blood, and silver pheasants.

The common pheasant landed in the United States in 1881, when the U.S. consul to China, Judge Owen Nickerson Denny, had twenty-six of the birds shipped from Asia to the sprawling grasslands near his home in Oregon's Willamette Valley. The birds thrived. Nine years later the first U.S. pheasant season opened in Oregon, and the take was some fifty thousand birds. They have since been introduced to forty states and they have readily adapted, for while they may look dandyish with their long regal tails, upright carriage, and array of colored feathers, they are tough little birds, equally at home on the beastly hot, rattlesnake-infested prairies of Texas as in the well-below-zero weather of Montana and the Dakotas.

Its brilliant plumage and noble appearance make the ring-necked pheasant an exotic-looking creature on the American landscape, especially as they dart across the muted gray-green high plains of northern Montana. There's no other wild bird out there that is anything like them. Cocks have a white neck ring and a brilliant red wattle against indigo and black or blue coloring. The feathers on their body are a gorgeous rainbow mosaic of gold, green, black, purple, white, copper, and brown. Pheasant tail feathers are extralong and come to a sharp point.

Even though I carry a twelve-gauge shotgun, hunting pheasants is a challenge, for the birds are wily, especially after they have been pursued by hunters and dogs for a few days. When the shot starts flying, the birds head for the thickest cover, or hide near a rancher's house and barn or under his tractor because they know these are safe zones. When they flush, they are low, fast fliers, cruising just above the ground at speeds of thirty or forty miles per hour, though only for short flights of several hundred yards.

I hunt on the vast swells of remote and sparsely populated prairie that stretch along Montana's border with Alberta. The region is called the High Line, after the single train track that was built through here in the late 1800s. It's a remote and largely forgotten place dotted with small ranching and farming towns in decline, with empty storefronts and abandoned homesteads.

Often I start where the silhouette of the Bears Paw Mountains sits like an illusion in the distance, a short drive from the tiny town of Chinook, Montana. Walking across the broad sweep of grassy prairie of northern Montana day after day is an immersion into a starkly beautiful landscape. On the days I hunt I become a predator, and the experience touches some deep and ancient part of my psyche, a calm, though vigilant, deeply felt energy, providing me with the stamina to hike mile after mile along creeks and down one-lane dirt roads, all but oblivious to distances covered or the hours passed, consumed only with thoughts about in which patch of chokecherry, cattails, or thick grass the birds might be hiding.

There is a transcendent aesthetic to the entire bird hunting experience. When I hunt I pay attention to my world very differently than in my usual life. I use my senses in ways I don't normally use them, listening for a rustle in the thick, brambly creek bottom, noting the direction of the wind, watching for the flash of brilliant colors as a cock runs into a grain field far ahead, remaining alert to rattlesnakes, and smelling damp grass as the frost melts late on October mornings.

Hunting behind a trained and passionate dog is central to the experience. It's exhilarating to wait for a bird to flush as an intensely focused, loudly sniffing canine disappears into a wall of grass as high as his head or a thick curtain of shuddering cattails to ferret out roosters too wise to fly and become a target until they no longer have a choice. The anticipation is primal for both the dog and me.

The experience is largely why wing shooting is big business. The game bird farm and shooting preserve industry places its value at nearly $2 billion a year in the United States alone, as people pay big bucks for a few days of pheasant hunting. The cost of three days at the John Burrell South Dakota pheasant lodge, for example, at the high end, runs $3,300. Shooting with the same company in England and Wales costs $2,000 to almost $4,000 per day.

Sustainable hunting for birds is a tool for conservation, as hunters work together to protect habitat and populations. The unbridled hunting of wild birds for food, however, is a top threat in many parts of the world. Between China and Korea, dozens of species of shorebirds migrate across the Yellow Sea, stopping on vast mudflats to rest and refuel for the remainder of their long journey. Development and climate change have taken their toll on the birds in these habitats, but the dagger through their heart is large-scale subsistence hunting with large monofilament nets that snag many thousands of flying birds. Among the species most at risk is the spoon-billed sandpiper, a gorgeous and rare little bird whose beak, as its name suggest, broadens dramatically at the tip. With just a hundred pairs left, its future is in doubt.

Birds can also serve as guides for hunters. When a hunter from the Boron tribe is ready to head out into the bush of southern Ethiopia and Kenya to gather a bit of honey, he emits a loud, sharp whistle through a clasped fist to summon his partner, the greater honeyguide. The birds don't always respond, but when they do, like a friendly character from a Disney cartoon, the gray and white bird shows up and hovers, ready to help out—for a cut of the action. Sometimes the bird initiates the search and flags down the human, giving out a *tir-tir-tirr* call and moving restlessly in the hunter's presence, like a dog begging to go for a walk. Prior to the hunt it has likely already located the bees—for researchers have seen the birds peering into the hives before dawn while the bees are fast asleep.

Either way, when it's time to fetch the sweet stuff, the bird flies back and perches near the tribesman, flashing its white tail. As the human hunter nears, the bird flies ahead again and again, signaling each time with its tail to show the way. Once the bird reaches the site of the honey, it changes its tune and gives out a soft "indication" call. When the hunter arrives, he fires up some bark to smoke the bees out of the hive, then splits it open with an ax or machete. Then he divvies up the find. The bird gets the wax, pupae, and larvae, and the hunter keeps the honey. Some believe the honeyguide must also be given a taste of the sweet stuff or, next time out, its retribution may lead a tribesman to a venomous snake or a lion.

Hunting is a way of survival for both animals and the aboriginal tribes around the world that still depend on it for sustenance, shelter, medicine, and warmth. I also eat the pheasants, sharp-tailed grouse, ducks, and other wild birds that I hunt and kill, but the real value to me is not their meat; it's the way they help me maintain a connection to wild nature—for a week or two a year they allow me to turn once again into a predator, with heightened senses, to experience a feral and largely forgotten part of myself and to live fully again on a land-scape, an experience that is all too rare in our modern world.

# CHAPTER 8

⌒∘⌒

# The Miracle of Guano

Guano, though no saint, works many miracles.

—PERUVIAN PROVERB

Bird guano makes the natural world go round. It's essential to a healthy planet, a powerful fertilizer that Mother Nature uses strategically to nourish and expand life around the globe. It's part of what scientists call ecosystem services. While all things that birds do for us are considered services—whether they give us models for building better airplanes, provide dinner, or serve as a window into the long-ago era of the dinosaurs—there is a subset of things birds do that help balance the natural world, from eating bugs and cleaning up disease-causing waste to bringing us detailed and otherwise unobtainable information about what's going on in the world. Of course birds are just one element of nature that supports humans; virtually everything from trees to soil to water all contribute greatly to our health and happiness. Birds, though, are in such great numbers and cover so much territory that they are able to bring us gifts unlike any other ecosystem service provider. And one of these important gifts is their poop.

Large deposits of seabird guano are found on islands around the world, but the largest by far was found on three small, white, crusty uninhabited islands in the cerulean waters of the Pacific Ocean, thirteen miles off the coast of central Peru. Things came together in just the right way to create this singular natural treasure. It is, for example, the good fortune of the Chincha Islands to sit smack in the middle of one of the world's richest marine ecosystems. Prevailing winds from the southeast push warm water away from the coast, which creates room for the upwelling of the Humboldt Current, nutrient-rich cold water from Antarctica. This in turn stimulates the exponential growth of the tiny organisms called plankton, a rich food source for massive schools of small fish, especially the anchoveta, a silvery fish that grows to about eight inches long. Eighty percent of the world's anchoveta swim near the Chincha Islands and are a magnet for giant flocks of seabirds, which gobble them up and then drop their load. The red, white, gray, and brown stripes of guano on the islands were two hundred feet thick or more in some places when the commercial miners first arrived. Despite the fact that the islands sit in the middle of the Pacific Ocean, they are arid, which is important to the high quality of the guano, for in moist environments the minerals leach back into the sea and the dung is poor.

The star of the guano show, named after its output, is the guanay cormorant, a two-foot-tall bird with a black back, a white belly, and red circles around its eyes. These birds number around two million, and breed along the rocky coast of Peru. The guanay cormorants make their ring-shaped ground nests out of desiccated guano, which led the native people who saw the glowing white rings at night to refer to this place as *quillairaca,* or "the moon's vagina."

The blue-footed booby also contributes to deposits here. Named for the Spanish word *bobo,* meaning "dunce," because they landed on ships at sea where they were easily killed by hungry sailors, boobies are among the most beautiful and unusual seabirds. Their oversize feet are a gorgeous powder blue that flashes as the boobies do their high-stepping waddle-walk and their elaborate slow-motion mating dance, during which they "skypoint"—throw their heads back and whistle. When the guts of these and other birds work their magic on the hordes of small fishes, the result is some great shit. And lots of it, thousands of tons per year. For thousands of years, millions of birds have been eating billions of fish and raining their poop down onto the rocks below.

Nitrogen and phosphorus are vital to things that grow, but until this mound was discovered, it was very rare to find them in such a pure state in such great quantities. The Incas recognized and honored the value of this *wanu*, or guano, and for centuries they mined it here and then carried it up into the mountains to apply it to their crops. It was so potent and precious that its use was restricted to the elite, and they made it a crime punishable by death to kill or disturb guano-creating seabirds.

The rest of the world discovered Peruvian guano in the nineteenth century, when the pioneering Prussian naturalist Alexander von Humboldt (for whom the Humboldt Current is named) set up his laboratory equipment in Callao, Peru. Von Humboldt was a polymath who studied natural phenomena around the world, everything from botany to geomagnetics to meteorology. When he arrived in Peru he was intrigued by how the native people revered guano. His research and writings on the subject provoked deep curiosity in England about this miracle substance, and the leading farmers of the day clamored to try it. The first twenty oak casks of guano landed on the docks of London in the 1840s. "A good guano is the most valuable of fertilizers," wrote Professor S. W. Johnson of Yale. Rich in phosphorus and nitrogen, Peruvian guano created one-third more yield for many crops, didn't have the foul odor of other manures, and cost far less to boot.

Guano became Peru's largest export and led to a giddy guano

era—the country's most prosperous time—that built fortunes and empires. The international mining company W. R. Grace, for example, was founded as a guano mining concern. In 1844, 29,000 tons were shipped to England from Peru; by 1860 that number was more than 350,000 tons.

On these barren, isolated islands in the middle of the Pacific, the clinking of pickaxes, the scraping of shovels, the coughing and hacking of nearly naked men in the sun, and the calls of many thousands of birds circling in the sky could be heard for decades. Dozens of freighters would line up in the heaving waves off the Chincha Islands, and barges creaked under the weight of oak casks filled with manure. It took three months to fill the hold of a single ship. One British naval officer counted a hundred large ships from eleven different countries loading simultaneously.

Extracting these riches took a steep human toll—the air in the filthy mines was rife with the clouds of lung-damaging dust particles—and in the 1800s the rate of death among the army deserters, convicts, and indentured Chinese laborers sent to work there was high. Workers were hard to find, and some mine owners resorted to "blackbirders"— slave ships that sailed from island to island in the Pacific looking for unwary natives to kidnap and force into working the dung heap. They would anchor off an island and entice the men to trade, and then, once they were on board and under arms, the ship would sail off to the mines.

Demand for seabird guano continued to soar, especially as it became a source of the nitrogen-rich saltpeter essential to the manufacture of explosives. It was so prized that in 1856 the United States passed the Guano Islands Act, which treated bird poop as a strategic material and authorized the Navy to seize any island rich in guano as long it was not under a foreign government's jurisdiction. More than a hundred islands came into U.S. possession this way.

The guano economy began to collapse in the 1870s as a new source of nitrate was discovered in Chile's Atacama Desert. And by the early twentieth century, synthetic nitrogen had been invented and the need for guano faded away. Meanwhile, as the chicken and hog industries grew after World War II, Peru made a decision to aggressively exploit the vast swarms of anchovetas by grinding them into fish meal for livestock feed. As a result, Peruvian seabird populations plummeted.

Still, the great guano era had changed the world. In a 2013 tome called *Guano and the Opening of the Pacific World,* University of Kansas professor Greg Cushman traced the skeins of guano's far-reaching global impacts. Until guano was discovered, manure from other animals and legume crops were the only ways to add nitrogen to farm soil, and they were in limited supply. The massive Peruvian deposit made vast quantities of nitrogen more easily available and lifted the limitations on crop production around the world. Because food was no longer restricted by what farmers could grow with a limited supply of nitrogen, Cushman argues, guano fueled the development of global industrial capitalism. Developed countries expected plantation crops and more meat, because turnips could now be grown year-round and in large numbers as cattle feed. Guano from bird-inhabited islands also fueled colonization, especially of New Zealand and Australia, because farmers were no longer limited by the nutrient-poor soil there. And a ready source of nitrates from guano made explosives much more available. Cushman also argues that Peruvian efforts to protect the seabirds and their riches provided fodder for the naturalist Aldo Leopold's thinking on conservation.

These days the Chincha Island deposits are staging a bit of a comeback. To capitalize on a surge of interest in sustainable, non-chemical ways to farm, Peru has rebirthed its guano industry, though on a small scale. Once again miners with pickaxes and shovels, wearing handkerchiefs over their noses and mouths or no masks at all, rise before dawn to break open slabs of petrified guano and scrape it into piles. They then pack it into 110-pound sacks, winch the sacks onto barges, and tow them to the mainland, about 23,000 tons per year. The guano is used by organic banana farmers and medical marijuana growers, but the bulk of today's crop is distributed to small organic farmers in Peru as a way to reinvigorate abused farmland. Today's guano mining is a little more enlightened than it was over a century ago, but not by much. Filling and moving more than a hundred sacks a day, workers make double Peru's minimum wage of about three hundred dollars a month. They work for eight consecutive months, and then take four months off to recuperate.

The Peruvian government works to protect the birds that still live there, because they realize the value of their droppings. When the miners leave, a guard stays behind to defend the flocks of big birds from poachers, who use bright lights to stun birds before they club them to death with sticks and take the meat to sell. Wooden frames have also been built to collect and protect the droppings.

The highest and best use of guano, though, the real gold for all of humankind, is that it is a highly effective system for the spreading of plant genetics. Birds are the FedEx deliverymen of the natural world. By flying around with seeds encased in a little package of potent fertilizer, untold numbers of birds spread life around the world every day, in ways that no other animal comes close to. Ecologists call birds "mobile resource linkers," a role that is crucial for maintaining ecosystem function and resilience. Not only do they deposit guano-encrusted seeds throughout the planet, but by eating the fish along the Peruvian coast and many other places, for instance, millions of seabirds transport phosphorus and nitrogen from the sea to the land, where these rich nutrients can be absorbed by a wide variety of terrestrial plants.

Cagan Sekercioglu is an associate professor of biology at the University of Utah in Salt Lake City. He's tall and intense, especially when it comes to the topic of birds and biodiversity. We met in his spacious office and lab atop one of the university's buildings, and then walked to a nearby bagel shop for lunch. Sekercioglu describes himself as a biophiliac—that is, he has a passionate interest in the earth's array of life, especially birds, and is pained by the assault on it. He's duty bound, he says, to help stem the great extinction now under way. He formed a nonprofit organization called KuzeyDoga

Society, or Northeastern Nature Society, to protect an unusually biodiversity-rich region by the same name in his home country. His work has earned him attention and accolades—the royal family awarded him England's highest honor for conservation work at a ceremony at Buckingham Palace, and he is one of the National Geographic Society's Emerging Explorers. He's also a deeply devoted birder, and he has trotted around the globe to gather more than seven thousand species on his life list. "Even if you just look for birds you'll see the best parts of the planet," he says. "Not just landscapes and biodiversity, but some of the last remaining interesting cultures."

To make the argument for protecting birds, Sekercioglu has devoted much of his time to researching and publicizing the essential things they do for us. If we won't protect biodiversity for its own sake, he says, we should do it for ours. As a rule, birds have not been viewed as very important providers of services—but that's only, it seems, because people haven't looked. The value of these services to humans globally, and to all of nature, is incalculable, he says. Yet because they are free, they're taken for granted. Sekercioglu has set out to change that. He's scoured the literature for studies that show how bird services benefit humans, has done some studies of his own, and wrote the definitive piece on the matter, a 26,000-word article titled "Ecological Significance of Bird Populations." He is one of the editors of a book called *Why Birds Matter,* a collection of research on the ecosystem services they provide.

For example, birds are pivotal for migrating plants, for dispersing seeds a long way from the parent tree in a nutrient-rich package. This bird transportation system is a matter of survival. If seeds simply drop off a tree onto the ground, the up-and-coming crowd of young plants must compete with siblings in the same plot for sunlight, water, and other resources, and many do not survive. "When you concentrate the seeds like that, it's a bonanza for seed predators," Sekercioglu says. "They are easy to find and then destroy all at once. It's putting all your eggs in one basket." That's not a problem, however, if the predators don't digest the seeds completely. When a flock of famished birds, or many flocks, gobble up a bunch of seeds and fly off in myriad directions, it changes the equation. A study showed that out of five hundred seeds dropped by a tree on a square meter of forest floor deep in the tropical forests of the Peruvian Amazon, only four became saplings. When birds grabbed those seeds and carried them off, however, more than 95 percent became trees. Because birds like to be in the light and saplings need light, this assisted migration was enormously important to the survival of trees. And it was even more important for rare tree species. Furthermore, as the climate changes, growing warmer or colder, plants need to migrate farther north or south to stay within their preferred range, and being carried inside a bird is one of the principal ways the seeds make the journey, Sekercioglu explains.

In November 1963, a new island pushed up out of the surface of the cold, storm-tossed Atlantic Ocean twenty miles off the southern coast of Iceland, the product of a volcanic eruption on the sea floor. It was named Surtsey, after a mythical Norse god of fire named Surtr, and it grew to the size of about a square mile before erosion from the pounding surf reduced it by half. The tiny speck was claimed by Iceland, and officials immediately declared it a nature reserve and laboratory. To preserve its ecological integrity, just a handful of scientists were allowed access to the island, and they watched closely as life colonized the barren rock.

In 1965 the first vascular plant established, called sea rocket, and two years later the first mosses and lichens floated to the island. A few years after the eruption, bird feet touched down on the island to live. The first birds were fulmars, gull-looking creatures, save for their tube-shaped nose. Large black-backed gull and kittiwake colonies arrived, and probably did more to encourage new life on Surtsey than any other birds by dint of their sheer numbers. They brought not only their rich fish-fueled guano, but an array of plant seeds. (Charles Darwin once germinated the seeds in bird poop and found that they represented a dozen different species of plants, including yew, holly, and raspberry.) The birds also carried seeds with the grasses they brought to Surtsey for nests, and other seeds came along stuck to their feet. They may even have brought the first earthworms. And as their eggs were raided and smashed, the yolk, albumen, and shells became the first layer of organic matter that formed soil. More than a half century later, fourteen bird species and some seventy species of plants—fields of waving grass, horsetails, ferns, and buttercups among them—call Surtsey home.

And if you love a spicy burrito, thank your feathered friends, for birds are a linchpin in the world's eclectic variety of spicy cuisine. In an ingenious partnership, chili plants make sure the birds get what they need—a nutritious food source—while the plants get their seeds spread far and wide. Capsaicin is the active molecule in spicy peppers that creates the heat we feel when it binds to pain receptors in the tongue and mouth, and they telegraph that burning sensation to the brain. On a human tongue, a solution of just ten parts per million creates a profound sense of burning. Nerve receptors in birds, though, don't register the searing heat of the capsaicin; even if they ingest a walloping twenty thousand parts per million, the birds merrily keep gobbling. Some bird fanciers, in fact, add hot chili powder to their feeders to deter squirrels and other raiders while allowing birds to keep eating.

Bird digestive systems also do the seeds of the chili pepper no harm, because many birds with a fondness for them have a rather short, straight digestive pipe through which food transits quickly. (Researchers have even found snails that had been eaten by a bird, passed through its digestive tract, and been cast out the other end, and were still alive.) Moreover, chili peppers can be plucked readily from the stem only when they are ripe, which means just those seeds ready to germinate leave the bush inside the bird.

Catching the eye of birds helps pepper plants to expand their range. If mammals ate chilies, they'd digest and destroy them or deposit them fairly close to the plant. Birds, however, drop the undigested seeds in faraway places. The cradle of chili pepper civilization is in Brazil, in a lowland region dubbed "the nuclear area" because it has the largest number of wild varieties of chilies in the world. It's believed that the first wild chili peppers sprang to life here and were spread across much of the Western Hemisphere by birds.

Birds shape the plant world in myriad other fascinating ways. Some two thousand species inadvertently play the middleman in sexual relations between plants. White-winged doves carry pollen between saguaro cactus flowers, and hummingbirds seek out the nectar of a variety of wildflowers and domestic crops; some hummers visit thousands of flowers in a day, and as they search for nectar they carry millions of tiny grains of pollen stuck to their heads, beaks, and feathers, swabbing flowers like flying paintbrushes.

There are places where birds and the pivotal roles they play have vanished. The extinction of the dodo, the three-foot-tall flightless bird of Mauritius, could well be the cause of the decline of the tambalacoque tree, also known as the dodo tree, which is highly valued for its timber. Some plants have just a single animal partner that plays a key role in its life, and in the case of the peach-tree-like tambalacoque, the dodo's digestive tract may have been vital in paving the way for the tree's successful growth. As the seeds moved through the bird's system, fruit pulp was scrubbed off, reducing the risk that bacteria and fungi would kill the seeds before they germinated. These days, botanists growing tambalacoque trees pass their seeds through wild turkeys or even gem polishers to roughen and clean them in preparation for planting.

Large enough flocks of birds can also be a factor in the weather. When millions of migratory seabirds arrive in the Arctic each summer, they bring a huge burst of guano. As bacteria go to work on the guano and break it down, they emit ammonia. As sulfuric acid and water combine with the ammonia in the atmosphere, bigger particles are formed, which become the nuclei for cloud droplets and thus create cooling clouds. The number of cloud droplets near seabird colonies can be 50 percent more than in those areas without birds. Smaller droplets are also formed, which reflect sunlight and contribute to the cooling effect. When seabirds change their migratory patterns or disappear, experts believe they may contribute to changes in Arctic weather.

Could birds that have disappeared someday make a return and reprise these ecological roles? A serious effort to make this happen is already under way.

There were once some six billion passenger pigeons in North America, perhaps 40 percent of all birds. It was common for a single flock of a million of these graceful, acorn-eating pigeons—depicted in paintings with their long, slender tails, peach-colored breasts, and red eyes—to descend into an oak forest and roost in the trees. As many as eight birds built their nest on a single branch, and the woods echoed with the sound of branches breaking and crashing to the ground under their weight. As the crowd of pigeons spread the processed acorns far and wide and deep—the forest floor would be covered with several inches to a foot of their droppings—their impact on a forest was robust. The famed naturalist Aldo Leopold called them "a biological storm . . . a feathered tempest that roared up and down and across the continent, sucking up the laden fruits of the forest and prairie, burning them in a traveling blast of life." As occurs with a forest fire, the tons of nitrogen-rich droppings killed grass and trees, but they also sowed the seeds for the explosion of a new, vibrant forest.

Market hunters traveled the country by train to find the birds' crowded roosts, and when they did, they shot as many as possible, salted them, and stuffed them into barrels. A hundred oak barrels a day were filled and shipped to food markets in New York alone, each barrel holding as many as six hundred birds. When the birds and their guano were wiped out, so was the periodic rejuvenation of oak, beech, and chestnut forests, which made the forests much less productive and less diverse.

Stewart Brand hopes one day to restore something of this forest dynamic. Long ago he created the Whole Earth Catalog, a large-format, visually rich book rooted in the counterculture of the 1960s, famous for its collection of innovative tools and ideas for back-to-the-landers. He's now seventy-eight and runs the Long Now Foundation near Berkeley, an organization that advocates long-term thinking. One of his projects, called Revive and Restore, is making an effort to bring back several extinct species, something he calls "de-extinction." The passenger pigeon is first on his list.

The band-tailed pigeon is a close living relative of the passenger pigeon, and it lives in large numbers along the West Coast, Brand explained to me. The two species branched off from a common ancestor a million years ago and divvied up the continent. The band-tail doesn't have the striking red eyes or peach-colored breast that its disappeared cousin did, but its DNA is 95 percent the same.

Experts from Harvard and the University of California, Santa Cruz, are working with Brand to sequence the genome of the band-tailed pigeon as well as ancient DNA from museum specimens of the passenger pigeon. They are scouring this biological bar coding to find the differences between the two—the snippet of DNA that represents the red eye, for example. Then this ancient DNA combined with band-tailed pigeon DNA will be injected into a band-tailed pigeon egg, editing it, in effect. "What you get," Brand told me, "is a degree of passenger-pigeon-ness," a hybrid between the two. "After you start to get a sense you are on track and the new genes are on track, you go through captive breeding for a long time," with a pair of the hybrids. After generations of breeding for passenger pigeon traits, the outcome should be something extremely close to the original. "It's in its early stages, with exploratory technologies, but the practicalities are getting more practical all the time," he says. A similar de-extinction effort successfully brought back an ibex in Spain, though the offspring lived only a few minutes.

"Once you have come up with a passenger pigeon, the project will turn to reintroducing it into the wild," says Brand—no small feat, since the nut-and-seed-producing forests it favored have been so degraded. Still, he says, it's worth a shot. "This was the mind-changing extinction that happened in American history," he says. "People knew the dodo had been wiped out, but that was on an island, and everyone assumed the continents were relatively safe. When the last pigeon died in 1914, it was a real shock. To undo that piece of destruction has a bit of redemption in it, and an apology." And perhaps a paradigm shift: the beginning of a future in which we can undo some of the gravest environmental damage we have done by bringing back the vanished and their role in the world.

# CHAPTER 9

## Nature's Cleanup Crew

And what did they do for us? They went to work, and
by thousands and tens of thousands, began to devour
them up.

—BRIGHAM YOUNG

After I interviewed Sekercioglu in his university office in Salt Lake
City, I drove downtown. Salt Lake City is the heart of the global empire of the Church of Jesus Christ of Latter Day Saints, also called
the Mormon Church, and buildings sacred to the sect are located in
Temple Square. As I walked around the meticulously kept grounds,
I stopped in front of the towering Gothic-style Assembly Hall and
looked up. A large pair of flying gulls cast in bronze, their wings
spread, sat atop a stone globe that rested on a tall stone pillar, an
homage to what is known as the Miracle of the Gulls.

In 1848, the first Mormon settlers were struggling to survive in

Utah's harsh desert. An abundant harvest meant surviving winter, and at first the fields looked bountiful. But as their crops were growing in the warm sun of early summer, a dark, ominous cloud appeared on the horizon above the foothills near the Great Salt Lake. Millions of crickets were headed in their direction. About the size of a man's thumb, these insects are ravenous and often devour whole fields, leaving nothing green in their wake. It looked as though the wheat, corn, and vegetables were doomed. Then on June 9, suddenly—and, many believe, miraculously—huge flocks of California gulls appeared on the horizon. For three weeks the pioneers watched as the heroic gulls gobbled crickets to fill their bellies, regurgitated them, and then ate more. "The sea gulls have come in large flocks," one witness wrote, "and sweep the crickets as they go; it seems the hand of the Lord is in our favor." The gulls saved the crops. The artist Mahonri Young, one of Brigham Young's 321 grandchildren, cast the sculpture of the insect-eating gulls that now sits in Temple Square.

Though some say the story of the Mormon crickets is apocryphal, birds are indeed big bug eaters. Nearly three-quarters of the ten thousand or so species of birds feed on insects, from ants and bark beetles to mosquitoes and flies, so the impact is enormous—on bugs both good and bad. One small study shows that migratory songbirds alone gobble three thousand to ten thousand tons of insects each day as they travel north from Mexico to the United States during spring migration.

China learned a lesson about the role of birds in pest control the hard way. In the postrevolutionary China of the 1950s, Chairman Mao Zedong ordered the countrywide elimination of four pests—mosquitoes, flies, rats, and sparrows, specifically the Eurasian tree sparrow, a diminutive bird with a white chest and belly, brown wings and back, and a black beard and mask. Chinese scientists had studied the matter of the sparrow and factored that each bird ate nearly four pounds of grain per year, so that every million sparrows eliminated would free up enough grain to feed sixty thousand people. In 1958, at five o'clock on a December morning, the Great Sparrow

Campaign began with the blaring of bugles, the screaming of whistles, and the banging of pots and pans with spoons and ladles. Shouting "Death to the sparrows!" millions of Chinese, whipped into an angry froth by a propaganda campaign, swarmed into the streets. "Chase them, hound them, scare them," officials exhorted, which, along with thousands of scarecrows and waving flags, kept the birds flying until they dropped dead from exhaustion. Soldiers gunned down birds in the large open areas of cemeteries, schools, and parks. Young men and women, including the Nanyang Girls Middle School Rifle Team, took to their weapons to wipe out the enemy sparrow. One zealous young man was singled out for his bravery in killing twenty thousand sparrows by climbing trees, smashing the birds' eggs, and strangling whole families in the nest. Millions of sparrows across China were successfully eliminated and, to celebrate, posters featuring heroic-looking boys and girls with slingshots and a clutch of dead sparrows were displayed all over the country.

When the corpses of the little birds were dissected, scientists were taken aback. Three-quarters of the sparrows' diet was insects, not grain. Chairman Mao halted the slaughter, but his edict came too late. Locusts and other insects the birds would have eaten were now free to devour grain and other crops, and the near extinction of the sparrow contributed to the great Chinese famine that began in 1958 and killed thirty million people.

There's a more recent lesson in how birds are important in helping maintain ecological balance. In the 1940s, the brown tree snake made its way to the thirty-mile-long Pacific island of Guam, perhaps stowed away on a military plane that landed when the island was strategically important in the Pacific theater of World War II. Since that first snake slithered off the plane, presumably carrying eggs, numbers of the snake in Guam—where they have no natural enemies—have grown wildly, and in some places there are three thousand brown tree snakes per square mile. The ten-foot-long reptiles crawl through the trees or lie in wait to eat birds, and they are damn excellent hunters. By the 1980s they had destroyed the bird population and eliminated ten out of twelve species—including the Micronesian kingfisher and Guam flycatcher. The remaining two species survived only in a small safe zone where snakes had been eliminated.

Without bird predators, spiders on the island have proliferated wildly, their population growing by up to forty times as many spiders found on similar islands. "There isn't another place in the world that has lost all of its insect-eating birds," says Haldre Rogers, an evolutionary biologist at Rice University who studies the ecological repercussions of the vanished birds. "You can't walk through the jungles of Guam without a stick in your hand to knock down the spiderwebs." And the jungle is eerily silent because of the missing songbirds. The absence of Guam's birds has also changed the makeup of trees in the forest because of the role birds play in distributing seeds.

As the world's birds decline and disappear, the financial toll from increasing numbers of insects will rise. On a coffee plantation in the Blue Mountains of Jamaica, researchers wondered how bad the damage from the coffee borer beetle would be without the Jamaican mango and red-billed streamertail, among other hummingbirds, to eat them. To find out, they built chicken-wire fences around coffee trees that would keep the birds out but allow tree-damaging insects in. Where birds could get at them, they gobbled up about half the beetles. The value of the hummers' bug-eating service was pegged at about 12 percent of the value of the crop, or at about $125 per acre.

In addition to bugs, birds eat critters that are much larger. Like cattle. And even people. And, believe it or not, this feasting also greatly supports the human world.

A plodding cow stumbles in the dust, then finally drops to the earth, and within minutes the nine-foot wingspan of a vulture casts a dark shadow onto the ground where the animal lies. Soon there is another silhouette, then another, and before long more than a dozen vultures soar overhead, as sure a sign of a death as the appearance of a scythe-wielding grim reaper. The vultures' enormous wingspan gives them incredible flying power and mobility, and they cover vast distances, even across countries, scouring the landscape for carcasses.

There are two main types of vultures, the Old World birds of Europe, Africa, and Asia, and those in the New World of North and South America. Old World vultures locate carcasses by sight, while some New World vultures find rotting flesh with a keen sense of smell. Once the birds land, it's not long before dozens are gathered around the feast; it's not unusual to see two hundred birds circling a single carcass. It's obvious why they have the name "vulture," derived from the Latin for "tearer." In as little as twenty minutes, a venue of vultures can rip apart and dispose of an eight-hundred-pound cow carcass. Though we might think of these big birds as distasteful and ghoulish—Darwin called them "disgusting"—they have evolved specifically to carry out the vital task of turning dead bodies safely into earth.

The vultures' bald domes keep bacteria from spreading as they plunge their heads into rotting remains, and their short, tough beaks can snap bone. "The tongue is very strongly grooved like an old-fashioned coal scuttle, and the edges are serrated like backwards pointing teeth," wrote vulture expert Peter Mundy in his book *The Vultures of Africa*. "No doubt the griffon vultures use their remarkable tongues to shovel food down the throat as fast as possible." Their naturally high body temperature and extraordinarily strong stomach acid allow them to eat just about anything with impunity. They are nature's garbage disposals, cleaning up the landscape and preventing the spread of tuberculosis, anthrax, and other diseases. No wonder the family name is Cathartidae, derived from the Greek root *kathartes,* which means the purifier.

There are some things, unfortunately, that even vultures can't eat. Two short decades ago three types of Old World vultures—the white-backed, slender-billed, and long-billed—were the most abundant raptors in India, as well as in nearby Pakistan and Nepal, all told some forty million of them. The ominous and ubiquitous hunch-shouldered birds were habitués of rooftops, temple walls, and spires, where they would wait for the drop of a fresh carcass. More than 80 percent of the population of India is Hindu, and so they are vegetarians, though many farmers keep cows or goats for their milk. India has more livestock than any other country except China, and little in the way of sanitation services, so when an animal dies it is very often left where it dropped. The cleanup of camel, cow, goat, and other carcasses has always been left to the vultures.

Biologists from the Bombay Natural History Society noted in 1984 that 353 breeding pairs of oriental white-backed vultures lived in Keoladeo National Park. By 1999 they had all disappeared. It was clear that a major die-off of vultures was under way, though no one knew why. By the early 2000s, vulture numbers in India, Pakistan, and Nepal had plummeted, from tens of millions to about eleven thousand—from superabundant to critically endangered in less than a decade. The collapse is "perhaps the most dramatic decline of a wild animal in history," Susan McGrath wrote in *Smithsonian* magazine. Wildlife biologists scratched their heads. What the heck was going on?

In 2000, Munir Virani, a Kenyan raptor biologist working for a conservation group called the Peregrine Fund, arrived in Pakistan to try to discover the possible cause for the stunning collapse. For three years, forensic toxicologists, the CSI of the bird world, took samples from dead vultures and sent tissue to labs in the United States. The tests revealed the culprit—something called diclofenac, a cheap, ibuprofen-like veterinary drug fed to domestic cattle and other livestock to ease sore, chapped udders. When the birds fed on a contaminated carcass and ingested the drug, it created a white, crystalline coating on their kidneys that led to gout, renal failure, and finally death. India eventually banned the manufacture of veterinary diclofenac, though it's available for human use and thus still widely used in livestock. The fate of the vultures remains very much in doubt.

The disappearance of vultures was only the first part of the trag-edy. As the decline of the avian garbage disposals rippled through the ecosystem, the number of cattle and other carcasses left to rot on the ground soared, since there weren't enough birds to eat them all. With more meat lying around, the number of feral dogs that roam the Indian landscape increased by a third—to nearly thirty million. More dogs means more rabies, and a much greater likelihood of human infections. Widespread distribution of a rabies vaccine in 2009 should have decreased India's sky-high death rate from rabies, but it didn't; rates increased instead, which experts attributed to fewer vultures. Rats feed on carcasses, too, and their numbers boomed, causing an outbreak of the plague. One study estimated that the loss of vultures in India cost a whopping 48,000 human lives and more than $34 billion.

The decline in vultures touched the spiritual lives of Indians as well, for the birds are woven throughout myths and traditions. Some Asian cultures have long held that the birds carry the souls of the dead to heaven, an ecosystem service you won't find much about in the scientific literature.

Several years ago an Indian woman, Dhun Baria, distributed to her neighbors grainy, ghoulish photos of human bodies rotting on a massive marble platform called the Tower of Silence, which sits in a decidedly upscale Mumbai neighborhood. The woman is a member of a small religious sect called the Parsis, a 3,500-year-old Zoroas-trian religion with some 200,000 members. She was trying to call attention to their plight. The earth, air, and water are sacred realms, the Parsis believe, and would be defiled if corpses were disposed of in them. So since before Christ walked the earth, they've held what they call sky burials. In Mumbai, the dead are wrapped in white muslin burial shrouds and placed in a circle on the Tower of Silence, an eighteen-foot-tall stone dhokma, where vultures land and, in an hour or less, consume them, purifying the bodies and releasing the spirits. As Indian vulture populations spiraled down, Baria entered the tower to see how her recently deceased mother's corpse had fared and was horrified to find her body and others piling up and rotting. "The staff told me everything," she recounted to a journalist for Agence France-Presse. "They said, 'Madam, it's a hell inside.' I was crying a whole day and whole night. It's the biggest mistake of my life that I have put my mother inside there. I thought I must help to change this system." She made leaflets with pictures of the mounds of corpses in the burial site to distribute to the Tower's upscale neighbors and warn them about what was going on. Eventually the distraught Parsis set up magnifying-glass-like solar panels to help speed the decomposition, but this system has not performed nearly as well as the flying "purifiers" once did.

Vultures in Asia will not be coming back anytime soon. In an attempt to restore the population, they are being raised in captivity and fed at open air "vulture restaurants," where humans provide them animal carcasses in the wild. But vultures are excruciatingly slow breeders. They lay just one egg per year, and it takes five years for the bird to reach breeding age. It doesn't help that the drug that helped wipe them out, though banned, is still widely available.

The huge decrease in the vulture population is not only happening in India. Globally, fourteen of twenty-three vulture species face extinction. Their complete disappearance in Africa, where they are widely poisoned and sold for traditional medicine, is a very real possibility. Poachers—elephant poachers in particular—have killed thousands of vultures because several dozen big birds circling a freshly killed carcass can lead authorities to the scene of their crime. They've taken to using pesticides—cheap and plentiful—which they pack into carcasses. "In East Africa we're seeing declines of sixty percent in vultures over the last two or three decades, but to be honest it's likely much faster than that," Darcy Ogada, a biologist with the Peregrine Fund who works in Kenya, told me. In one incident alone in northern Namibia, she said, six hundred vultures were killed, birds that had flocked from other countries to feed on a pesticide-laced elephant carcass.

The vultures of Asia have all but disappeared while few people were paying attention and the vultures of Africa are in trouble as well. The fate of these birds, and the fate of the ecosystem they are essential to, and which is in turn essential to many people, still hangs in the balance. It's a teachable moment in how we need to understand the world around us: not only those things that are beautiful and give us pleasure, but even those that seem, as Charles Darwin said of the vulture, disgusting. Sometimes beauty is simply waiting to be understood.

# PART III

Discovering Ourselves
Through Birds

# CHAPTER 10

❦

# Bird Brain, Human Brain

*I am a brain, Watson. The rest of me is a mere appendix.*

—SHERLOCK HOLMES IN ARTHUR CONAN DOYLE'S
"THE ADVENTURE OF THE MAZARIN STONE"

To become a taxi driver in London one needs something cabbies call "the Knowledge." Acquiring it is a grueling, mind-bending ordeal. London's ancient warren of streets lacks the gridlike order of cities such as New York and Paris, and so for several years prospective cabdrivers crisscross the tangled skein of its arteries on mopeds, up one street and down another, until all twenty-five thousand are committed, to the extent possible, to memory. Then the would-be drivers must take a torturous exam. If they pass, they are allowed to become London cabdrivers.

What does such hypermemorization do to the brain? Studies of seed-caching birds such as the Clark's nutcracker and the chickadee show that feats of memory substantially add to the pile of gray matter in the same way a bodybuilder adds muscle—the more they use it, the bigger and better it gets. Every fall, nickel-gray nutcrackers with black wings and black tail feathers cache as many as 33,000 nuts in 2,500 different locations. Then they unerringly recover some two-thirds of the protein-rich seeds more than a year later. Their keen memory is driven by the exigencies of survival—because of the ecological niche they occupy, they are far more reliant on stored food than other birds.

The brain's hippocampus—its name derives from the Greek word for seahorse, which it resembles—is central to the creation of short- and long-term memories and spatial perception in humans and birds alike. Scientists looking into the nutcracker's incredible feats of seed-caching memory, one of the best in the bird world, found that the hippocampus in these birds is far more robust than in birds that don't cache.

Is the same thing true in humans? It seems like it. Based on the bird work, researchers from the University College London tested cabdrivers who had acquired the Knowledge and found that those who drove the hackney carriage performed better on memory tests and had a far more robust rear hippocampus than peers who did not drive cabs. And the longer someone drove a cab, the plumper their hippocampus. Researchers wondered if the results could be due to a process of self-selection—did those who became drivers have larger hippocampi and better memories to begin with? In a follow-up four-year study, seventy-nine would-be cabdrivers were selected who had all performed similarly on memory tests, and all had their brain imaged to measure the size of their hippocampi. Only thirty-nine successfully acquired the Knowledge and went on to become cabdrivers. The twenty who didn't pass the test remained in the study.

Testing showed that four years later the memories of those who had passed the test had improved beyond those who failed—even though their memories had been similar at the outset—and imaging showed that their hippocampi had grown larger. Just like the nutcracker, a cabdriver's feats of memorization are also achieved by the demands of economic survival.

The drivers themselves, though, were puzzled by the findings. "I never noticed part of my brain growing," David Cohen, a member of the London Cab Drivers Club, told the BBC.

The fact that exercising our memories and performing other cognitive skills makes the brain bigger, stronger, and better able to carry out its tasks is a major finding in neuroscience. And our feathered friends have played a role. While birds provide a way for us to look back on how dinosaurs lived, they also allow us to peer in a very different direction—deeply into the human brain. They are an important and unique model system, one that allows us to understand the extraordinary plasticity of our own brains, the toll that stress exacts on us, and possible cures for a variety of neurological problems.

Harvey Karten, eighty-six, is a legend in the rarefied field of evolutionary comparative neurobiology. A professor emeritus at the University of California, San Diego, who still does research, and a member of the National Academy of Sciences, he began his career with studies of the human brain in 1963 and was on his way to becoming a psychiatrist when, during his residency in Colorado, he heard a lecture by Konrad Lorenz, aka Father Goose, the ornithologist famous for imprinting baby ducks and becoming their surrogate father. "I was stunned," Karten recalls, "that ethology, the study of animal behavior, had so many parallels to what we study as psychologists. I thought to myself, 'We don't have all the bias and assumptions of knowledge about birds and the bird brain and the relation to its behavior that we do about human brains.'" So began his interest in bird brains and behavior as a model for the human condition. "They tell us things about people we couldn't otherwise study," he says.

Karten has probed the gray matter of owls, pigeons, and a variety of chicks in the hope of better understanding the corresponding features of the mammal brain, and when, and in which creatures, those features originated. He identified in birds, for example, the optic tectum, a large, vital, and "beautifully layered structure," he says, as if he were describing a jewel in his hands. It governs how creatures, including humans, move in space, how they shift the direction of their gaze and respond to what they see. "And it's a uniquely sensitive mechanism for detecting very small objects moving at high speeds, like a mosquito in the periphery." Karten also has done much work on the auditory cortex in birds, the sliver of brain that processes what they—and we—hear. "It's amazing, it's exactly the same as mammals," he says. And it's why birds have such excellent hearing.

For a long time, bird brains were thought to be radically different from—and far inferior to—mammal brains. "The human brain is constructed like a layer cake," says Nicola Clayton, a professor of comparative cognition in the Department of Psychology at the University of Cambridge. "A bird brain is built like a fruitcake." Researchers like Karten, who study a range of brains, see in a human or chimp or even a giraffe or mouse, brain structures that are fairly similar across the board. The neocortex, the critical top two-thirds of the mammal brain, for example, has six layers, or sheets, of neurons that grew so thick during the course of evolution that craggy folds and fissures evolved to pack all that gray matter into the small skull space, similar to the way you stuff a down jacket into a small backpack. The neocortex is the hallmark of the modern mammal, wherein complex communications and abstract thinking, the expression of emotion, and many elements of personality are carried out. When you learn new facts about birds and memorize them, you can thank your neocortex.

Birds, though, don't have one. They do, however, have an equivalent, called the pallium. It looks very different—it's very smooth, not craggy and fissured. Because of the stark differences between bird and people brains, it was assumed for more than a century that birds must be stupid—"birdbrains" with only instinctual drives and no ability to assess and respond to the world the way mammals do. It wasn't until the last decade that scientists realized that most of the thinking about bird brains was wrong. Birds are far smarter and their brains are far more sophisticated than anyone imagined. To reflect this new knowledge, a global team of twenty-nine scientists from six countries formed a group called the Avian Brain Nomenclature Consortium in 2005 to rename the parts of the bird brain. And in classic fashion, as researchers realized how intelligent birds really are, they wanted to find out more about them. Fascinating stories about bird smarts began to emerge.

- Pigeons have taken the Eysenck IQ test, which was designed for people. They were shown three keys in a row with an asymmetrical symbol on the center key. The same symbol is shown on one of the keys to the side, while a mirror image is displayed on the other key. To get food, pigeons had to peck the side key that matched the center key as the symbols to the side were rotated to different angles. During the test, pigeons performed better than humans routinely do.
- Some birds can count. Alex, the famous African gray parrot, got as high as six. In the 1970s, biologist Pamela Egremont watched as Chinese fishermen on the Li River used cormorants to catch fish. The fishermen place a neck ring on the birds, which cinches their throats tightly so they can't swallow, and they are then trained to return with a fish in their mouth. The birds are allowed, however, to eat every eighth fish. When they return to the fishermen after the seventh fish, Egremont wrote in a journal article, they "stubbornly refuse to move again until their neck ring is loosened" so they can eat the eighth fish as usual. "They ignore an order to dive and even resist a rough push or a knock, sitting glum or motionless on their perches. One is forced to conclude that these highly intelligent birds can count up to seven."
- Crows have a remarkably robust memory, and can remember a human face that has done them wrong for years. University of Washington corvid researcher John Marzluff and two students wore rubber masks to test crows that live on campus. A caveman

mask was deemed "dangerous" and a mask of Dick Cheney was labeled "neutral." Students in the dangerous mask trapped and banded seven crows. Over the next several months, volunteers donned the two types of masks on campus, this time walking prescribed routes and not bothering the crows. Yet the crows remembered the "dangerous" faces. In a display of anger common to crows, they vocally scolded people in a caveman mask much more loudly and harshly than they did before they were trapped, even when the mask was disguised with a hat or worn upside down. The neutral mask provoked almost no reaction. And despite no further trapping, the scolding of the angry crows when they continue to see the caveman mask has only grown more intense over the last few years. Crow memories are so good that several years ago the U.S. Army was looking into spy birds that could be trained to memorize faces of missing soldiers and targets such as Osama bin Laden, and then go out and search for them and report back.

Early in his career, Karten thought the bird brain might be a completely different organ than the human brain. "Birds had solved the problems of motion control, of depth control, of balance and three-dimensionality, of auditory recognition, of optic flow, so the question I asked was 'Did they really solve it in a totally different way?' If they had, that would be fascinating, and discovering a different kind of brain that did the same things human brains did would be like discovering a whole new galaxy. But as it turns out, bird, mammal, and reptile brains are variations on the same theme."

In 1859 a geologist and, ironically, an evangelical Christian, John William Dawson, made one of the great discoveries of evolution. While rummaging about in the red rock and coal seams of the Joggins Fossil Cliffs of Nova Scotia in Canada, Dawson uncovered the fossil of a lizard about as long as his hand, which he named *Hylonomus lyelli*. It's now believed that this critter is the last common ancestor of birds, lizards, and mammals, including humans. For the last 360 million years or so, reptile, human, and bird brains have evolved in very different ways, but they are all based on the same fundamental model.

The realization that bird brains could be a surrogate for understanding the human brain sparked a revolution in neuroscience—especially because many birds have their own language, whose sophistication is unmatched in the animal world. One of the earliest—and so far most earth-shaking—secrets to be coaxed from the bird brain was discovered in the 1980s by the Argentine neuroscientist Fernando Nottebohm, who was doing research at the Laboratory of Animal Behavior in Millbrook, New York, part of Rockefeller University. As a child growing up on the pampas of Argentina, he was a bird-watcher and loved to listen to their songs. Nottebohm wondered how the tiny brain of a chickadee could remember where the seeds it had stashed were hidden and how canaries learned their songs every year. New neurons, perhaps? Conventional wisdom at the time held that the number of neurons bestowed on a creature at birth was its only allotment in life—the brain could never create new ones. But because of previous studies, Nottebohm knew that the size of different structures in a bird's brain changes with its environment. Just like the nutcracker, a wild chickadee has many more neurons in its memory center in the hippocampus than a caged one, for example, because if it doesn't remember where it has hidden its seeds, it could starve to death. Still, at the time, the notion that brain structures changed over the course of a life by adding neurons was, Nottebohm told *The New Yorker,* "considered the view of a lunatic. People in my own lab begged me to stop. I saw the pity in their eyes. They were saying, 'Fernando has lost it completely.'"

He persisted, though, injecting his canaries—he used canaries because they learned new songs each year—with a special dye for a week. A month later he examined their brains. As he peered intently into his microscope at one of the little birds' brains, he was shocked to see that the tiny organ had created new cells. The number of neurons in the brain region that control singing in male canaries soars during the mating season, he discovered, driven by a surge in hormones that create the new storage for songs the birds are learning, which they use to woo potential mates. "Every bird, young or old, was producing thousands of them each day," Nottebohm said. Come summer, when mating is finished and songs are no longer needed, the number of neurons plummets.

Science does not often immediately reward those who upset the established order, and Nottebohm's work was first ignored and then attacked. Top brain experts vehemently denied that any such thing could be possible and insisted that his results must have been a mistake. "Read my lips: No new neurons," said Pasko Rakic, an eminent primate brain researcher at Yale who bitterly opposed Nottebohm's discovery and is credited with single-handedly holding back the acceptance of the finding for years. Unfortunately for Rakic and fortunately for progress, additional research propelled by Nottebohm's findings showed that other animals—and humans—are capable of neurogenesis as well.

"That is the holy grail for us," said Arturo Alvarez-Buylla, one of Nottebohm's former students, who now researches neurogenesis at the University of California, San Francisco. The implications are enormous. "What we are talking about is teaching the brain to repair itself with its own cells. It's not going to be a simple task. It's a type of magic really, but eventually I think it's going to be possible. And for that we should thank Fernando and his birds."

There's good reason for the bird brain to be so dynamic: flight. Just as their bones are full of air and their feathers extraordinarily light to enable them to fly, the bird brain needs to be as small as possible in order not to weigh the creature down, so new neurons are added for caching or crooning when they are needed, and shed when they are not. The same thing goes for birds preparing to migrate— they not only add weight to their body mass to provide energy for the trip, they also add new neurons to the brain to aid navigation during migration. And the further they migrate the more neurons they add.

The brain's astounding plasticity works the other way as well—that is, it shrinks and loses function under duress. Birds have been enlisted in the effort to better understand this phenomenon, and it's all because of their singing.

Ondi Crino, a researcher at the University of Montana, opens a door to a closet-size room in the back of the same Flight Lab in Missoula, Montana, where Bret Tobalske does his hummingbird work. Several zebra finches and a half dozen diamond doves flap nervously around our heads until they land on some branches and the door frame molding, anxiously looking down at us. As we move, they flutter and fly about, above us. "They're cute, aren't they?" Crino says, and while, yes, they are cute, what she is really interested in is not their looks, but their central nervous system, which is key to her work. Specifically, Crino is studying something called the HPA axis, which stands for hypothalamic-pituitary-adrenal, a complex and vital physiological system that is also found in all mammals, including humans. It governs the fight-or-flight response, almost instantaneously activated by exposure to fearsome events—for birds, that could be when a hawk soars above them, for example, or when the chilling eyes of a slithering viper that suddenly invades the nest appear. It's a system that keeps birds and many other living things, including us, alive and out of harm's way. Stress is a constant in the bird world. "Young birds are like Chicken McNuggets for predators," Crino says. So how, she wonders, does that stress play into the ecology of the birds? How does it affect their breeding? Their family life? The secret lies in the song of the finch.

Crino has made an effort to understand a few different animal species' nervous systems by their vocalizations. She has also researched singing mice, for example. White-crowned sparrows and canaries are other go-to animals for many researchers, and bluebirds are used in stress work because you can build a bluebird house and, simply by removing the roof, peer down and study the young ones as they grow. But the colorful gray zebra finches, with an orange beak, a slash of black and blue feathers, and a distinctive black-and-white teardrop under their eyes, are the most popular bird research species, the avian equivalent of the white lab rat. In the wild, the birds live in the grasslands of central Australia, Indonesia, and elsewhere, and their singing is loud and active. Aspects of their song have been studied since the 1950s.

The zebra finch learns one song of between four and eight notes in its first ninety days, and the song doesn't change much over its lifetime. Their music is not particularly appealing—it has been likened to the squawk from a mechanical cat or the sound of a squeeze toy. But what the bird lacks in the beauty of its song, it makes up for in its ability to be easily bred and kept in captivity with minimal care.

Crino's investigation into how stress affects a bird's life helps us unpack the ways in which stress impacts humans—which is just about every way—including the digestive system, the immune system, moods and emotions, sexuality, and energy levels. Stress plays an outsize role in anxiety, borderline personality disorder, depression, insomnia, and a host of other psychological problems. In fact, it is likely the single biggest preventable factor in human health, so understanding how it works is vital. "We can look at these mechanisms in finches in ways we could never look at them in humans," Crino says.

Crino uses a couple of hundred male finches in just one of her experiments, buying them at a pet store for about fifteen dollars each. She divides them into two populations—one the group being experimented on, to which she feeds a stress chemical called corticosterone, which is mixed with peanut oil. That amps up their stressful feelings. To the other, the control group, she feeds plain peanut oil.

One measure of Crino's studies comes a month later, when she listens to hear if the chemically stressed birds have changed their tune. Over several years of research, she has found that being the wild version of a Chicken McNugget does indeed hamper the bird's cognitive abilities and memory, particularly its ability to learn its signature song. And because the song it sings dictates a zebra finch's choice of a mate, stress can stand in the way of a bird's very survival. "Stress decreases the size of the neural structures responsible for song learning and production," says Crino, as she sits on a couch sipping tea in her lab. And because an ambitious, full-throated song is their primary come-on for the girls, "stressed males are less sexy and less attractive."

These gathering storm clouds of stress, however, have a silver lining. Male birds with a poor song, at least by lady zebra finch standards, are apparently so thrilled to finally land a mate that once they choose a female, they devote themselves to her, lavishing attention and food, provisioning the babies more than less stressed birds do, and cheating with neighboring lady birds less (among researchers, the polite term for philandering is "extra-pair copulation"). "The bottom line," says Crino, "is that contrary to what we might think, developmental stress is not always bad. It can change behavior and physiology in ways that can be adaptive."

A good deal of light has been cast on the human condition because of this kind of work, and it could lead to major treatments for serious afflictions. Humans are capable of neurogenesis, though only in a few brain regions. Birds can do it robustly throughout the brain. If we learn the mechanisms in birds that control this regeneration, a human brain could possibly be coaxed to create new neurons in specific regions, the way bird brains do; perhaps it could even usher in a new era of therapy for stroke, trauma, Alzheimer's, Parkinson's, and other brain ailments. That's why around the world scientists keep cages of finches and canaries, studying human brain physiology via the bird brain, answering basic wiring questions and investigating such things as how neurons respond to learning by creating networks and why the human nervous system declines as we age. The bizarre state of semi-limbo that is bird sleep might help us understand human sleep disorders. And some research into bird memory, especially in the Clark's nutcracker, shows promise as a treatment for Alzheimer's and dementia. At age seventy-six, Fernando Nottebohm is still working with his birds, looking for insights into Huntington's disease, a neurodegenerative motor disorder.

Just as watching a bird run and learn to fly opens a window on the evolution of flight, peering into a bird brain adds to our knowledge about how it evolved from the dinosaur brain. But it may go back even further and lead perhaps to the source brain from which dinosaur, mammal, and bird brains all sprang. "That's the biggest chocolate chip cookie ever made," at least for people like him, Karten told me. "Where did the brain come from? When did it first arrive? No one knows yet, but birds give us insight."

We do know, though, that "the basic system was in place long before mammals and dinosaurs came along," he added. Studies on lampreys, an eel-like jawless fish that has been around for 535 million years, much longer than dinosaurs, show that they have all the major components—the neurochemistry, the interactions, the connections—and have the same physiology at a single-cell level as birds and mammals.

There is another big bird brain question, which will not be answered anytime soon. Right now scientists have teased apart various structures and understood the brain's function, yet they are still a very long way from understanding how all these pieces bind together in real time—similar to the way the various instruments in an orchestra play a symphony—to create the exquisite movements of a bird in flight. Karten's observation is much the same as Tobalske's and Dial's: Birds are a wonder and a humbling inspiration. "Watch an Arctic tern when it is fishing," Karten tells me. "It's flying along, it drops its head and looks through Snell's window"—a physics term for a perceptual porthole that allows birds to see into the water without distortion—"and it then does a dive and goes into the water, grabs the fish, and goes up and out. It's working in a beautiful visual motor vestibular platform we cannot even begin to approach and understand. Yet they do that every moment of their existence."

# CHAPTER 11

⤙⤙⤙ ⌾ ⤙⤙⤙

# The Surprisingly Astute Minds
# of Ravens and Crows

With other animals you can usually throw out 90 percent
of the stories you hear about them as exaggerations. With
ravens, it's the opposite.

—MARK PAVELKA, U.S. FISH AND WILDLIFE SERVICE,
IN *MIND OF THE RAVEN*

I sat in my car along a snowy road in Yellowstone National Park's
stunningly untamed Lamar Valley, my binoculars fixed on several
wolves as they loped across an expansive, sun-drenched prairie. Several black birds circled in the cloudless blue sky above, their wings
outstretched and not flapping, riding an unseen current. Crowds of
people often line the road here to watch wolves hunt, but few pay
much attention to the flocks of ravens that travel with them, though
they might if they knew of the Machiavellian tactics and extraordinary intelligence on display, especially in the orgy of feeding that

surrounds a fresh kill. Ravens on a kill are an exhibition of one of the premier animal intelligences in the world.

Wolves and ravens have some kind of understanding: They are partners, hunting and eating side by side—which is why the naturalist Bernd Heinrich calls ravens "wolf birds." They depend on wolves to kill elk, deer, and other animals and rip their carcasses open, things that ravens can't do on their own. Wolves, in turn, often rely on ravens to spot game from their highly mobile aerial vantage point and call attention to it. When ravens land at an animal carcass and are unable to open it, they call out loudly and repeatedly until a wolf shows up to tear it open.

When an elk is brought down by a pack of wolves, ravens show up at the kill almost immediately—in no longer than two minutes in Yellowstone. The average number of the black birds that turn up here in the park for a kill is 28; the record is 135. They gorge alongside the ravenous wolves, jockeying for position, bobbing and weaving and dashing in to rip off a piece of bloody meat, as the best of the bonanza won't last long. Ravens must be smart and nimble enough to get a share without getting killed or injured. That's why, when they are young, or in the lull between kills, they often joust with wolves, in a kind of daring horseplay, to test the animals' response and learn which wolves might bite, how far the birds need to jump to stay away from the razorlike teeth, and how to distract a wolf long enough to snag some meat. "The situation can change in a millisecond, from one individual wolf to another," Heinrich explained to me. "So you have to learn reactions and be able to figure things out and anticipate from moment to moment. That requires quite a bit of intelligence." Heinrich speaks with authority. He lived intimately with wild and captive ravens for many years while he studied them, and wrote the highly acclaimed 1999 book *Mind of the Raven*.

I drove down to Yellowstone, just a few hours from my home in Helena, Montana, to find out more about these birds, and the stories are legion. In the town of West Yellowstone, at the park's west entrance, a small tourist town where they measure snow in feet instead of inches, dozens of people suit up daily for a snowmobile trip into Old Faithful's steaming geyser basin. In summer the ravens mooch off the hordes of tourists, wandering through parking lots and picnic areas. Come winter, though, scraps are scarce, and so the birds have figured out the art of unzipping zippers and unsnapping bungee cords in order to open packs fastened to the backs of snowmobiles and help themselves to food inside. The Park Service posts warnings: WE *WILL* ROB YOU reads a sign above a mug shot of a glaring raven. "They work as a team to open any compartment on your snowmobile," the Park Service grimly warns. "They will take your food, your keys and even your wallet, if you let them." Then what? Will the smart-alecky corvid run up large sums on your credit card?

While ravens most often scavenge, they are sometimes hunters themselves. In early May 2001, a Yellowstone ornithologist, Terence McEneaney, witnessed an unusual and ruthless act of raven predation. On a quiet spring day, more than a hundred eared grebes, black-and-brown diving birds with striking red eyes and tall crests, accidentally landed on an iced-over Yellowstone Lake, thinking it was water. Once they plopped down, the grebes couldn't take off again, because they couldn't find purchase with their lobed feet on the slippery surface. A raven flew out to one of the stranded grebes and killed it by stabbing it again and again with its beak. Then it killed another. And another. Three more ravens joined in. "After killing the ninety-second grebe the ravens began to dismantle their prey, caching the grebe remains in the snow along the shore of Yellowstone Lake," the ornithologist wrote. "At the end of the day all one hundred and forty-one eared grebes had been killed."

The common raven—aptly named, for they are ubiquitous in the Northern Hemisphere around the world—is a member of the Corvidae family, which includes crows, jays, nutcrackers, rooks, magpies, and a few others. Ravens are the largest of the corvids, with a wingspan that can reach nearly five feet. The color of jet, they have the ominous and brooding appearance of an undertaker. Their kin, the crow, look similar but are smaller, and while ravens travel often in pairs, crows like the company of flocks. Ravens are great fliers and their aerial acrobatics are something to behold. They somersault, roll, and even fly upside down. Ravens can live to be forty years old, and they keep the same mate much of their life. Their most famous attribute, though, is their extreme intelligence, and they and their fellow corvids are the best evidence in the animal kingdom for the existence of a mind.

While the architecture of their brain is far more sophisticated than we once knew, the fact that they are "smart" birds may only be part of the ravens' story. It's one thing to poke the right key on a screen test or count to seven, but experts think some birds—especially the corvids—go beyond being merely clever and have a much richer and more complex awareness than we know. They can reason, strategize, use logic, understand cause and effect, and practice Machiavellian tactics by deceiving and manipulating other birds. They may even have the most human trait of all, emotions. That is a subject of much contention, however.

Some of the best evidence that ravens have a mind of their own comes from two of the world's top investigators of raven intellect—Heinrich, now seventy-seven and a professor emeritus at the University of Vermont, and his former postdoctoral student Thomas Bugnyar, forty-seven, a researcher at the University of Vienna, one of the premier institutions for the study of avian cognition in the world, who studies, among other things, how the social interactions of ravens shape their intelligence. Ravens are so wise, Bugnyar told me, that in a host of ways their behaviors mirror primate behaviors, including human primates. That's why investigating the complex mental behaviors taking place among a flock of ravens not only gives us a glimpse into the evolution of the black bird's extraordinary cognitive capabilities but also serves as a model to study the evolution of the human mind, which was driven by similar demands for survival: the need to navigate a socially complex, often strife-torn world to battle with others for food, sex, and power.

Wolves are not the only animal that has a close working relationship with ravens. For a long time, humans have woven them into their lives as well. Ravens may hold more symbolic meaning for us than any other bird. For cultures around the Northern Hemisphere, they play the role of muse, companion, portent, seer, an omen of bad luck and foreboding, as well as a source of inspiration. They are paradoxically associated with the darkest aspects of human life, as well as the most illuminated.

Like owls, ravens are widely seen as harbingers of death in many cultures. Swedes consider them the ghosts of murdered people, Germans the ghosts of the damned. In Edgar Allan Poe's famous eponymous poem, the ominous bird is a symbol of the author's descent into a dark cellar of madness. Perhaps because they are carrion eaters, they are believed by some cultures to be mediators between the seen and unseen worlds. They are central in the mythology of a wide range of Native American tribes and play a starring role in the cultures of coastal native tribes of the northwestern United States and western Canada, including the Haida tribe, who believe that the selfish and mischievous raven teaches us lessons about human nature. On the other hand, the Haida also view the raven as a deity; the tribe's creation stories describe the bird as an animal spirit that created land for people to live on, and who also stole light from the gods in order to illuminate the world.

And for centuries the raven's spiritual prowess has served as an indispensable, high-maintenance totem for one of the great countries of the modern world. Six good-luck ravens live as pampered guests of the queen of England on the manicured emerald lawns of the Tower of London, with a seventh raven in reserve. The corvid crew is overseen by a long line of Raven Masters, bedecked in regal red-and-black uniforms with matching top hats of sorts, well-paid employees of the crown who feed and care for the birds. As long as the ravens remain at the Tower, tradition holds, the United Kingdom will not fall to a foreign power. The queen's minions take no chances: Each bird has a clipped wing to keep it from leaving.

Ravens and their crow cousins have for centuries been seen as important auguries in human cultures around the northern world, from Tibetan Buddhists to the Athapaskan people of Alaska. In ancient Rome augurs were powerful priests who were schooled in interpreting the behaviors of ravens, crows, and other birds as a method to divine the future. Raven augury considers several elements, including a flock's different social classes, numbers, their sundry calls, their behaviors, and their location in the sky at specific times—collectively called "taking the auspices." During the "first watch," from six until nine in the morning, for instance, a crow arriving from the southeast meant an enemy was approaching, while a crow from the southwest at sunset was a sign that the wishes of one's heart would be fulfilled.

For centuries, we have valued the intelligence of ravens and crows. In an Aesop fable called "The Crow and the Pitcher," a thirsty crow finds a pitcher with water at a level too low to wet his beak. So he waddles over to some pebbles, picks them up with his beak, and drops them, one by one, into the pitcher. As he does, the water slowly rises to a level high enough to allow the crow to slake his thirst. This story is not far-fetched. This is just one of the things this canny bird is able to do.

On a snowy, cold-to-the-bone day in February, at a tiny, bustling café in Montreal, I chatted with Dr. Louis Lefebvre, a professor of animal behavior at McGill University who specializes in the study of bird intelligence and who made global headlines a few years ago with his bird IQ test. Over a steaming shot of espresso, Lefebvre, with salt-and-pepper hair and a long gray scarf wrapped around his neck, explained why ravens and other corvids are the smartest members of the bird world. He spoke especially of a whip-smart species called the New Caledonian crow that comes from a small tropical island far off the coast of New Zealand.

Lefebvre's favored way to measure bird IQ is to look at the novel ways birds find food, which is when they are sure to strut their smartest, most innovative stuff. Ornithologists have long recorded unusual field observations of birds and published them as short notes in birding journals. Lefebvre combed through seventy-five years' worth of anecdotes of feeding behavior among birds—more than twenty-three hundred of them—and assigned each account a score based on how many innovations the birds showed in relation to how intensely that species was studied. (A species on which hundreds of studies have been done, for example, is far more likely to be seen doing something new than a species that has been little studied.) The most innovative feeders? Lefebvre identified the brown skua, a seabird that butts in between seal pups to suckle nursing seal mothers. And he singled out the vulture. During the war in Rhodesia, hunch-shouldered vultures would gather near a minefield to wait patiently for unwary gazelles to wander by and get blown up. "It gave them a meal that was already ground up," Lefebvre explained. This kind of novelty indicates superior intelligence, he says, and these two events scored the vulture and brown skua high on the IQ scale.

No other family of birds, Lefebvre told me, can compete with the corvids when it comes to intelligence. In his study, the birds in the genus *Corvus* were five times as innovative as an average bird, such as the cuckoo, and two members of the Corvidae, crows and ravens, were eight times as innovative, distinguishing them as the intellectual elite of the bird world. Why are they so smart? There are three reasons. First, they have a huge pallium (the bird equivalent of the critical neocortex) relative to their body size. Second, they rear their young for more than a year, and more nurturing means less stress. And third, they live in a large, busy, and complex raven society.

It depends on how you define it, but the most intelligent corvid could be the New Caledonian crow, which is also in the running for the smartest animal on the planet. The research world realized this in 2002, thanks to a crow named Betty. Betty was plucked from the jungle near Yate, New Caledonia, a tiny French archipelago that once served as a penal colony, some four hundred miles off the coast of New Zealand. In March 2000 she was shipped to Oxford University in England, where she was welcomed by Alex Kacelnik, a behavioral ecologist who studies tool use in animals. Lefebvre and his team couldn't find innovative feeding anecdotes for the New Caledonian crows, because they live in a remote part of the world and are not well studied. But Kacelnik's laboratory tests provided convincing evidence for their smarts.

Betty and a fellow New Caledonian crow named Abel shared a room in a lab. During a filmed experiment in 2002, Betty and Abel were offered the choice between a short piece of either hooked or straight wire—both novel items to the crows—to grab with their beak and use to fetch a small bucket containing pieces of their favorite food, chopped pig heart. Kacelnek wanted to see if the crows knew enough to pick the hooked piece of wire to raise the bucket up a narrow, clear plastic pipe, the only way to access it. When Abel quickly absconded with the hooked tool, Betty was out of luck. But she didn't mope; she immediately and adroitly stuck the straight wire into a crack in a plastic tray and bent it into a crude hook. Researchers watched in amazement as she nimbly reached down with her newly crafted tool and raised the bucket so she could tuck in to the savory meal. Some animals use tools, but in a simple way—chimps dip a stalk of grass into an anthill and pull out a few ants and eat them, for example. But the idea of a bird modifying a novel object into a tool "is virtually unheard of" in the animal world, Kacelnik says.

"No other animal has demonstrated that high an order of intellect," Lefebvre told me. Betty repeated the experiment many times, showing it wasn't a fluke. And later research showed that crows understand cause and effect.

New Caledonian crows also exhibit something called "meta tool use," which means they not only use tools, they use them to fetch other tools to retrieve food. Gorillas and apes do it too, but it's otherwise an alien concept in the animal world. Kalcelnik's lab crows,

Luigi, Icarus, and Gypsy, all picked up and used a short stick to fetch a longer stick in a cage because it was needed to obtain a meaty morsel. It's a significant step, because it exemplifies behavior that is not going to directly provide a reward.

After that the crows got even more meta. One crow pulled up a string in order to untie a short stick tied to it, and then used the short stick to fetch a long stick, and the long stick to fetch the food, showing that they think, in the words of one researcher, "three chess moves ahead." The world of bird researchers again was stunned. It is one of the reasons crow expert John Marzluff calls crows "flying monkeys."

Perhaps Aesop's famous pebble-dropping crow was a New Caledonian. In one New Zealand study, a tiny bit of meat floated on a cork, too far down in a tube for the crow to reach. Given the choice between stone-size objects that floated and those that sank, the crows chose the sinking ones 90 percent of the time—the kind that could displace the water and float the cork high enough for them to reach the meat. This rudimentary understanding of fluid dynamics, researchers say, is equivalent to the workings of the mind of a seven-year-old child.

New Caledonian crows are just as smart in the wild as they are in the lab. In the dense, tropical evergreen forest and savannah that is their home turf, they're hard to observe on the go, so Kalcelnik's team fitted eighteen of them with tiny half-ounce cameras on their tail feathers as they went about their daily business. The corvid cam showed that these crows made and wielded tools constantly in their everyday life, whittling twigs and leaves into useful shapes to herd and capture ants and using long stems of grass to fish for insects. They seemed to particularly like prodding an inaccessible grub with a stick, and when the irritated insect responded by biting the stick, the crows would pull it out with the grub attached and eat it. The researchers discovered that tool-making styles vary on different parts of the island, suggesting a range of crow cultures. And just recently researchers discovered another tool-using bird, the Hawaiian crow, which is nearly extinct—just a hundred or so individuals remain in captivity in San Diego. Researchers believe that because both birds evolved on remote tropical islands where predators are few, they had more time to develop tool-using skills.

Why do the crows of New Caledonia have an edge on their corvid brethren when it comes to certain tasks? First, the bird's beak is unusually straight, "more like a human opposable thumb than the standard corvid beak," said Russell Gray, who studies the birds in Auckland. That means it's easier for them to grasp a tool. Its eyes are also closer together than the eyes of other crows, which means that a New Caledonian has a better view of the tool it's using. And its brain is proportionally bigger, more so even than those of other corvids, especially the front, the section that governs fine motor skills and learning. What would these birds be able to do, you have to wonder, if they had arms and hands like a chimp?

"Imagine ten million years ago, when God was deciding which animal to elevate, to become human," said Lefebvre as he spun for me these coffee-fueled tales of smart birds. "It was the great ape, but it could just as easily have been the corvid." Still, he said, the big question of whether animals have a mind complete with emotions and the ability to reason is, to Lefebvre—a behavioral scientist—unanswerable. "Birds can produce innovative behavior, and we can measure that, but I don't have a way to study whether they have a mind."

Not everyone feels that the question has no answer. Theory of Mind (ToM) is a type of awareness that some researchers say describes the highest form of consciousness. Simply put, it holds that a mind has an awareness of the belief, intentions, desires, and emotions in both itself and other creatures. According to this understanding, the essence of having a mind is both knowing and caring what others think of you. Humans are the only species with all the elements of ToM, many believe. Some researchers think, though, that ravens and other corvids may qualify for admission into this exclusive club.

A sophisticated language is one hallmark of a mind, and the verbose and brassy ravens certainly have one. They have a complex array of verbal calls, as well as names for one another. These calls are categorized variously as a gurgling croak, a shrill alarm, and a harsh grating call and, based on when they issue the sounds and how they differ, there are thirty-three types. Ravens are also famously great mimics, known to make the gurgling and swooshing sounds of toilets flushing in public restrooms and to tease road builders by mimicking the sound of exploding dynamite. They use a series of gestures as well, whether eye contact or ruffling wings, though these signs have only begun to be deciphered. Sometimes, ravens will pick up stones or twigs and wave them about in order to gain the attention of a fellow raven. This kind of gesturing was thought to be something only humans and other primates do.

Another element of an advanced mind is the ability to follow someone's gaze. If I look at a box of crackers, will a bird turn to see what I'm looking at? Most children develop gaze-following abilities by about eighteen months. Heinrich and Bugnyar both carried out a number of experiments testing the ability of ravens to gaze-follow. One at a time, several birds were set on a perch on one side of a room, separated from the room's other side by a barrier. An experimenter sat three feet in front of the side of the barrier and looked in different directions, including behind the barrier. In every instance the ravens followed the gaze of the experimenter. In fact, the birds not only followed his gaze, but sometimes when he looked where they couldn't see, they hopped down from their perch to waddle around the barrier, or sat on top to peer curiously over and have a look for themselves.

Self-awareness is yet another major indicator of the presence of mind. How conscious is an animal of itself? There's a *Far Side* cartoon that depicts an angry parakeet looking at its reflection in a mirror and, believing it's another parakeet, is ready to do battle. A fellow parakeet, though, holds the agitated one back and says, "Whoa, back off, Bobby Joe, that's just your own reflection!" If that combative parakeet was "self-aware," Bobby Joe might have recognized that the reflection was himself and chilled on his own. I often wonder this about my dog, Maggie, as she trots along with the family on a hike. Is she aware that she is a dog, that she is different from people in some fundamental way, or does she simply think she is one of a pack of the same type of animal?

The mirror test for self-awareness is as simple as research gets. Affix a brightly colored sticker to a baby's forehead and plop it down in front of a mirror and it won't react to that thing on its head, it will just continue to gaze at its reflection in the mirror. Test the same kid at about age two, though, and the toddler has developed self-awareness—it reaches up to touch the sticker, wondering "Hey, what is that weird thing on my head?" The baby, experts say, has developed a sense of self.

Dogs don't pass the mirror test—sorry, Maggie—though dolphins and elephants do. Ravens, surprisingly perhaps, do not, at least by this standard, have a sense of self. Another member of the Corvidae family, however, does—the European magpie. German researchers put the black-and-white birds in front of a mirror and placed a bright red or yellow sticker under their beak. The magpies responded to their altered reflection by touching the sticker with their foot or beak.

The most powerful indicator of the presence of consciousness may be the ability to mentally time-travel, a talent also known as *chronesthesia*. The term refers to the ability of a mind to remember images of events from its past and to use those memories to imagine possible future scenarios. Children can do this by the age of about four. Having the ability to mentally re-experience the past, and to imagine future scenarios in order to decide how to behave in the present, is very different from operating on instinct, as most animals do. If we make lunch for tomorrow after tonight's dinner, we aren't doing it because we're hungry but because we recall that we get hungry at midday at the office, so we plan ahead.

Mental time travel (MTT) is something that we do all the time, even if we don't realize it, and it is the source of our walloping success as a species, because it allows us to adapt to complex and changing environments. The great apes, considered one of the smartest species in the animal world, do not time travel mentally, and so they are limited by their "current drive states"—slaves, in essence, to a perpetual now. Caching seeds for future use, which the Clark's nutcracker does, is seen by some scientists as MTT, because based on past experience, it's aware that it will be hungry in a few months. Critics say it's not proof positive, though, since their stashing could be driven by instinct.

Nicky Clayton, a professor of comparative psychology at the University of Cambridge, disagrees with the naysayers. She created an experiment that refuted one part of the "it's only instinct" claim by giving lab scrub jays two kinds of food: moth larvae and peanuts. The jays love the larvae when they are fresh, but if they are stale, they prefer peanuts. The birds hid both of their favorite foods and were then taken away from their cages for four hours. When they were returned, they dug up the still fresh larvae. If they were brought back five days later, though, the birds dug up the peanuts instead, because they remembered that the larvae go bad after a few days.

Another Clayton experiment established the future half of MTT. Scrub jays were kept in three adjoining compartments for six days. Each morning, for two hours, jays were shut into one of two rooms—in one room there was nothing to eat, and in the other, there were pine nuts that were powdered so the birds could eat, but not cache, them. For the rest of the day the scrub jays moved among all three of the rooms. On the seventh day the powdered pine nuts were replaced with real ones—and the birds stashed three times as many in the room that had no breakfast. They were planning for the future. It was a breakthrough; many now think it's possible that birds are able to time-travel in their minds.

The biggest mind-related question about birds, and indeed about all animals, is whether they experience emotions. Lefebvre argues that we can't know. But there is disagreement on the subject.

Gisela Kaplan, an Australian researcher, believes that birds are very much emotional creatures, but they just aren't very good at displaying it, at least in human terms. "Our mouths are extremely important in all the nonverbal communication we have amongst each other," she says. "We express countless emotions just by mouths alone. But a bird has a hard, inexpressive beak, no teeth to flash, and eyes that have no whites." Still, she says, they do emote. "The entire head, and indeed the entire body, has very complex musculature, and birds can move feathers in a way that is used very effectively for communication." In a bird called a galah, for example, ruffled neck feathers might mean the bird is angry, and happiness can be expressed by lifting the crest. And many bird species purr when they are content, just like cats.

After decades of living and working with ravens, Heinrich is in agreement with Kaplan. "Sensations and emotions regulate behavior," he wrote in *Mind of the Raven*. "They serve to motivate and guide behavior when there is no immediate reward or logic to guide us. Since ravens have long-term mates, I suspect they fall in love like we do, simply because some kind of internal reward is required to maintain a long-term pair bond."

Still, Lefebvre argues that these are inferences, and that there is no convincing empirical evidence that animals feel emotion, nor will there be, since feelings are subjective and beyond the reach of science. Whether they purr, grin, or mate for life may merely be some type of instinct. Even PET scans on crows that show that their brain lights up in the same emotional areas of the brain as humans don't mean a thing. "Does your dog feel guilty when you yell at him for chewing up your shoe?" he asks. "It might seem so, but it might be that the dog is simply responding to your aggression. We interpret it as guilt, but we don't know if the dog feels guilty. It might just be a face that says 'Don't hit me.' We have no idea whether the brain creates the same internal properties in a bird as a human. We don't have access to their internal emotions."

There's a lot at stake in answers to these questions. How would we rewrite the ethics of a new relationship with birds if we accept the fact that they are conscious? If birds do have their very own mini-mind, should we still eat them? For now the answer will come from philosophers, not scientists.

Several times a month, Thomas Bugnyar journeys from Vienna to the montane forests and meadows near the village of Grunau in the Austrian Alps, home of the Konrad Lorenz Research Station, which was established in 1973 by Lorenz to study the evolution of animal intelligence. Knowing how incredibly smart birds are, Bugnyar says, raises the question of how they got to be that way. And in investigating that question among the big black birds here, he also hopes to shed some light on how humans became so smart.

Three things are believed to drive superior raven smarts: doting parents, a large brain relative to body size, and the fact that they evolved in a large and complex society. For six years, Bugnyar has meticulously observed and recorded the social dynamics of a flock of nearly three hundred wild ravens. His work concentrates on how the flocks live together, what the costs and benefits of living in a large animal society are, and how certain interactions among ravens in daily life may have sculpted their advanced mind.

Bugnyar pursues something called the *social complexity hypothesis,* which holds that the driving force in the development of a large brain and intellect—and humans and ravens are among the creatures with the largest brain size relative to body—was the need to anticipate, respond to, and manipulate others in the interest of survival. In raven culture that means negotiating and cooperating with other ravens in the flock for food and companionship and creating alliances to gain social status. And it also means deceiving and manipulating other birds.

The theory is that the practicing of these cognitive skills caused the raven brain to grow more robust and larger, and as it did, their abilities were further enhanced. This goes back to the lessons of the

Clark's nutcracker and the London cabdrivers. Bugnyar is not just interested in how these things happen in a lifetime but rather how they happen over the course of evolution.

Two other elements of social life play a pivotal role in raven—and human—mental evolution, Bugnyar believes. One is something called *fission-fusion dynamics*. Simply put, it's how often a bird leaves one group of companions and forms, or joins, another group. "Today in the morning, I leave my family, split up with my private group, and go to work with others," said Bugnyar. "Later I go to play basketball or soccer, and in the evening I go to have a beer with friends. It's important for me to learn about them and their relations and to figure out the best ways to use their relations for me," he said. "It seems to be a critical system." Critical in that the more complex and challenging the social environment, and the need to form, use, and maintain social bonds to gain advantage, the more the brain and intellect develop.

Fission-fusion is very much how ravens live, as opposed to starlings, say, who are almost always in a flock, or eagles, who mostly live solo or in pairs. Ravens roost together by the dozens at night, where a lively information exchange takes place, as well as playful behavior and grooming. During the day they fly out into the world, traveling alone, in pairs, or in small groups. They come together on an elk, deer, or flattened rabbit kill, and then splinter again when the meat is gone.

But knowing others isn't enough—it's using them to your advantage. And so politics is a big part of raven society. A key item in the raven toolkit is "Machiavellian intelligence," the ability to strategize what the competition is doing and to respond for your own benefit. Ravens form alliances, for example. "Say two ravens fight and one gets beat up," says Bugnyar. "The victim tries to get away. He flies off and hides in the trees. Within a few minutes another raven, which was not involved in the fighting business, comes over, and as he tries to approach the victim, who is still shaking and presumably nervous, he vocalizes very nicely, touches him, and starts grooming him."

Consoled, the victim flies off with his ally to rejoin the others. "Ravens not only form alliances, they try to prevent others from forming alliances to prevent them from becoming powerful as well," says Bugnyar. For example, they intervene when a rival raven is grooming another and take over the grooming themselves.

Ravens with multiple allies gain status and become even more powerful, in order to reap the rewards. "As soon as there is a restricted resource—food or a nesting site or shelter, whatever—those of high status get it," says Bugnyar. "It's an immediate benefit. But there also might be a delayed one, and a very important one: better access to breeding territories, so they can raise young, which would determine their fitness." Those that become powerful through alliances, in other words, are the ones whose genetics are most likely to be perpetuated. Evolution selects for the most politically skilled birds.

Food caching and pilfering play a major role in raven smarts, around which sophisticated cognitive behaviors grow. When ravens watch another raven cache food, they remember as many as twenty-five places where those birds have hidden their tasty morsels, and can recall those spots with high accuracy for twenty-four hours. Here, too, is where we see the theories of Machiavelli play out, in something called tactical deception. Ravens conceal information about their stash by being self-aware enough to know when other birds are looking at them and discerning whether the other birds know that they are hiding their food. They know which birds have food of their own and which don't, and therefore who's a threat. "They were able to put themselves in the other's shoes," says Bugnyar. Such behavior is one proof of a Theory of Mind.

In a lab study, Bugnyar used two of his captive birds, Munin, a dominant bird, and Hugin, the subordinate, named after the Norse god Odin's mythical birds. Prying off the lids of color-coded film containers to get at the cheese stuffed inside was a snap for Hugin, but he could rarely get out more than a piece or two before Munin, who was not as good at removing lids, bullied his way in and took over. Then things got interesting. Hugin feinted—he hopped over to some similar-looking but empty containers, pried off the lids, and pretended to voraciously gobble invisible food. When Munin came over to assume ownership of that food and was distracted, Hugin returned to the filled containers and resumed eating. Once Munin found out that the trick was on him, he threw a tantrum, hopping angrily about the cage, squawking and tossing food and empty containers about. "Ravens are very, very good at selectively withholding information or changing information or providing different types of information to particular individuals, depending on the situation," Bugnyar says. "They control their intentions and do particular things only in private. This is the first step toward Machiavellian politics."

Here, then, is one possible scenario for how the superior raven intelligence might have come to be. The birds evolved in the demanding environment alongside packs of snarling, ravenous wolves to be able to steal pieces of a rapidly disappearing carcass, evading the wolves in the middle of their meat lust, learning the right moves to avoid injury, and memorizing personalities—which wolf is fast and ornery, which ones are slow and accommodating. Ravens would grab a piece of elk loin or liver, fly out and cache it far enough away and in a hidden enough place not to be seen by other ravens, then they'd hurry to get back as the carcass rapidly disappeared. They had to control their attention, looking around to see who might have seen them hide their food, assess the threat by placing themselves in that bird's position, and then decide whether to relocate the morsel.

Those birds with superior cognitive skills—better memory, better attention, faster learning speed—who could deceive, manipulate, and befriend others, and make the most effective use of alliances, thrived over time, and their genes were passed on.

I asked Heinrich what the most valuable message is that the wolf birds hold for us. "That the animal world is far more intelligent and aware than we have presumed," he replied. "We are finding that capabilities that we have prided ourselves on and used as a rationale for our superiority apply to other animals as well. If we look at the intelligence, skills, capabilities, and fabulous behaviors, then we have to see the intrinsic value of these creatures. A lot of people might not be able to accept that we are not the ultimate crown of creation, but that seems to be the case."

Perhaps the most important gift the raven gives us is the gift of humility. They teach us that humans are not as exceptional as we might think, that we are not the only ones on the planet with higher levels of thought.

And of course our ignorance is in part due to the fact that these discoveries about bird minds, preliminary as they are, are based on only those questions we have thought to ask. Remember what the physicist Werner Heisenberg said: "We don't see nature, but nature that responds to our method of questioning." What about those things we haven't thought of asking? Rupert Sheldrake has long advocated that science needs to consider the nonmaterial realm, including the mind, and has long asked questions others don't. When he found a talking parrot named N'kisi that seemed to know what its owner was thinking, he decided to investigate. In a videotaped experiment he separated the parrot and its owner, Aimee Morgana. He showed Morgana a series of flash cards with scenes and she simply looked at them. N'kisi, in another room, would spontaneously say what his owner was seeing. N'kisi scored very well, and "the odds against this result being due to chance were more than 2,000 to one," Sheldrake wrote. But because the score recorded only hits and misses, the depth of the hits was not included. As Aimee was shown a picture of a car with a driver's head sticking out, for example, the bird said "Uh-oh, careful, you put your head out." Sheldrake believes the bird is indeed able to sense what its owner is thinking, an ability that he says may be widespread among birds and other animals.

This might sound bonkers in the worldview most of us currently hold. Many of the world's cultures, though, see all creatures in the world—all life—as conscious and believe there is a deep emotional connection between humans and the rest of nature that goes far beyond what we know today. This theory that all of nature is conscious, not just people, is called *panpsychism,* and up until a century or so ago, it was the dominant view. It seems to be making something of a comeback now, in part because of how much we are learning about the brains of birds.

It would help, of course, if birds could tell us their side of things. And while that may seem like a wild-eyed dream, some scientists are trying to do just that.

# CHAPTER 12

# The Secret Language of Birds

Birds scream at the top of their lungs in a horrified hellish
rage every morning at daybreak to warn us all of the truth
but sadly we don't speak bird.

—KURT COBAIN

In Hans Christian Andersen's fairy tale "The Nightingale," the
shrouded specter of death hovers over the gravely ill emperor of
China, poised to take his soul at any moment, when a plucky little
nightingale appears on the limb of a tree outside the royal bedroom.
"She had heard of the emperor's illness, and was therefore come to
sing to him of hope and trust," Andersen wrote. "And as she sang,
the shadows grew paler and paler; the blood in the emperor's vein
flowed more rapidly, and gave life to his weak limbs; and even Death
himself listened and said 'go on little nightingale, go on.'" Moved

by the bird's glorious song, Death takes a holiday and the emperor recovers.

It's easy to understand why Andersen chose the nightingale. Its song is one of the bird world's sweetest and most beloved; an older male—and in the bird world, the singers are almost exclusively male—may have more than two hundred fifty variations in his repertoire. And it's one of the handful of birds that sing at night.

People have been intrigued and enchanted by the complex sounds that come out of tiny birds since the beginning of human time. And as the tools to study birdsong have vastly improved, the fascination has taken on new dimensions.

Erich Jarvis stands out in the world of birdsong research. An African American from Harlem, he was raised in a gritty neighborhood by a single mother and his grandparents. His father, who was mentally ill and homeless and lived in caves in New York City parks, was murdered in 1989. The crime was never solved. Tall and lithe, Jarvis was on track to become a professional dancer—he studied ballet at Manhattan's renowned High School for the Performing Arts and was offered a position at the school of the Alvin Ailey American Dance Theater—when science lured him off the stage to a lab where he felt he could still be creative, but have more impact on the world. He is the only birdsong researcher I know who is attempting to transform a bird that cannot sing into one that does.

Jarvis, a former student of Fernando Nottebohm, is an associate professor of neurobiology at Rockefeller University in New York and was the leader of the consortium that renamed the architecture of the bird brain. These days he dances only as a pastime—"mostly salsa," he says—and instead spends most of his time researching how birds learn to sing, and what that process can tell us about how the human brain learns. He has spent years deconstructing the song circuits of the bird brain to identify the tiniest elements of song at the genetic and cellular levels. By unpacking song learning at these most basic levels, he is casting a great deal of light on understanding the details of how the human brain learns, whether there's a way to enhance it, or, if there is damage to the circuits, how it might be healed.

Zebra finches are the species of choice in Jarvis's research. Wild birds' brains are also used by scientists to study song learning. Call it extreme bird-watching: Jarvis sometimes goes into the field to gather up hummingbird brains, for instance, luring the birds to a feeder with sugar water. "They'll find the food source, and in the morning, as part of the dawn chorus, they will sing next to it," he says, which is how they lay claim to territory. Their singing activates a messenger molecule, and if their brain is removed and examined quickly enough, within half an hour—as humanely as possible, Jarvis says—he can find chemical traces of the song in the pathways along which those molecules traveled. Then he can measure changes in their brain activated by the birds' singing.

Jarvis has some big dreams about applying his rarefied bird brain knowledge. His project has the potential to transform birds that have never sung a note in their life, such as pigeons, into virtuosos by engineering brand-new circuitry in their brain. Pigeons have the hardware to sing—a syrinx, the tiny structure just above the lungs that is the source of bird vocalization—as evidenced by their cooing, but they lack the software—neurons in the brain that would generate more sophisticated tunes. "I've rebranded myself as a neuroengineer," he says. "If I can figure out how to induce a vocal learning circuit in the brain of a species like a pigeon and get it to learn how to sing like a songbird or imitate like a parrot, why, that would be my holy grail!"

The area of the brain that would be responsible for allowing a bird to develop song lies in the forebrain, which in a pigeon is about the size of a grape. Injecting new genes would, in essence, fertilize the song circuitry and create new axons in cells that would then connect to the motor neurons that control the fine movements of the syrinx.

If the technique proves successful—a big *if*, with plenty of techni-
cal hurdles to overcome—it may lead to a new era in repairing the
human brain. "If we can figure it out in birds, we can figure out how
to similarly repair circuits damaged in stroke and trauma in people,"
Jarvis says, or discover new drugs that help people regain speech
after a stroke, say, or find a cure for stuttering, a brain-based afflic-
tion that also occurs in some birds. The technique might also be able
to patch up the malfunctioning circuitry of the autistic brain. "I am
not sure I can do it in my lifetime, but I am going to try," Jarvis says
optimistically. "Maybe someday," he adds with a laugh, "we could
even engineer brain circuits for dancing."

In the gray twilight of a windless dawn one morning in Oregon's
high desert, I watched several dozen male greater sage grouse strut
their stuff. They are the dandies of the prairie bird world, wearing a
giant boalike vest of white feathers and a broad spread fan of brown-
and-white tail feathers. The air was thick with sexual tension, set to
a soundtrack of their clucking and bubbling calls, along with a mysterious popping sound. The calls came from their syrinx, I knew, but
I was not sure of the source of the popping. Finally, after glassing the
birds in the growing light, I noticed two yellow, balloonlike air sacs
emerging from their chests. During the dance, the cocks inflated the
brightly colored sacs and slapped them together emphatically to
make a sound that apparently mesmerizes the lady birds.

This sage grouse symphony is one of the bird world's most dramatic courting rituals. It takes place on thousands of *leks*—small
patches of grassy ground that are the main areas where sage birds
congregate—across eleven western states each spring. Feathers fly as
the males, who weigh up to seven pounds, clash to sort out their hi-
erarchy, sometimes physically, but mostly with their elaborate displays. The aim of the rooster as he brashly sidles up to a hen is to
catch her attention and impress her enough with his display to allow
him to plant his seed. If the hen likes what she sees and hears, she

turns her rear end toward the male, and, in a flurry of thrumming wings, the deed is done in a matter of seconds. To the victor go the spoils: The top cock mates with as many as three-quarters of the females.

What are these randy sage grouse saying with their twilight calls and slapping air sacs? A wide range of researchers are trying to decipher what birds are saying to one another and what they might be saying to us, especially songbirds. Intrigued that an animal so small can sing so sweetly and elaborately and communicate with its calls, researchers have made birdsong one of the most researched subjects in animal behavior.

And this research is not just about birds. Figuring out what they are saying may be key to understanding the animal world at large. "Even though I like birds and I study birds, I don't think there's anything particularly special about birds," says Mike Webster, director of the Cornell Lab of Ornithology's Macaulay Library of Natural Sounds, which has cataloged and archived thousands of recorded birdsongs. "But what birds are saying to each other might be fundamental in some ways to what whales are saying to each other. Or frogs. Or other mammals. Including humans." If we can learn to speak bird, "the most elaborate acoustic communication systems in the animal kingdom," says Webster, we may be able to communicate with our fellow species, or at least understand what they are trying to tell us.

The ten thousand or so species of the world's birds make just two categories of sounds: calls and songs. They communicate in other ways as well—through facial expressions, a range of eye movements, ruffling the feathers on their head, or simply rustling their wings or unfurling their tail feathers—but songs and calls seem to be the most favored methods. Calls are usually short and simple, while songs are longer and more complex. Calls are, for example, the caw of a raven, the cluck of a chicken, the quack of a duck, the screech of an owl, the honk of a goose, or the shriek of a hawk. Not all birdcalls are emitted by the syrinx; many are mechanical. Storks, for example, make loud clattering noises with their bills; hummingbirds emit a shrieking whistle with their tail feathers as a call to love when they dive during courtship; woodpeckers drum out their signature tattoo on a tree; and then, of course, there is the whacking of sage grouse man-boobs.

Some forty-five hundred of the ten thousand or so bird species in the world burst into some kind of song, and it is the Passeriformes, or birds whose four-toed feet are built for perching—three in the back, one in the front for a good grip—whose songs are most developed, especially the songbirds, which include thrushes, canaries, and warblers. In most species the songs come prewired; in just a few— parrots, hummingbirds, and songbirds—their vocalizations are learned. The birds that learn to sing their song from their parents as babies are the ones that most interest scientists like Jarvis, for vocal learning is very rare in the animal word. While elephants, bats, and the cetaceans—whales and dolphins—also learn to express themselves with voice, their abilities are nowhere near as sophisticated as those of birds. "If you raise a bird in isolation, it grows up and doesn't sound anything like its species," says Steve Nowicki, an animal communication researcher at Duke University. "In fact it hardly sounds like a bird at all. If you raise baby birds in the lab, like a parent, feeding them every half hour from dawn till dusk, and play them songs, they can demonstrate with remarkable accuracy the specific songs you are playing that they have learned."

That's why in the Vogelsberg (literally, Bird Mountain) region of nineteenth-century Germany, foresters snatched young whistling bullfinches from their nests and taught the affectionate little guys to sing by whistling to them. They taught them not birdsongs but instead folk tunes and more complicated musical pieces, including Chopin's "Thou Art So Like a Flower." Some of the birds learned three different songs. It's all the more remarkable because bullfinches don't have much of a song of their own. The birds became a fashion of the time, an expensive one, and everyone from Queen Victoria to Tsar Nicholas II owned a whistling bullfinch.

Song learning in young nestlings has two phases. The *sensory phase* in the first weeks of life comes from hearing their father sing. After these tracks are laid down, nestlings move on to the *sensorimotor phase,* where they try out their own song using what they learned from Dad's tune. They first vocalize a quiet "subsong," bird nonsense and babble, really, just like the nonsense syllables of a human infant. With constant practice they progress to recognizable singing, but still not fully formed, something called "plastic song." After two or three months, though, the bird has his own serious musical chops. And while a birdsong may seem fairly simple, when researchers slow it down they find that it is incredibly intricate and complex.

Each singing bird has a repertoire, which consists of different versions of a song, though some birds may have only a single song. One-fifth of singing birds have five or more songs. The North American brown thrasher has the largest repertoire of any bird, with as many as three thousand songs in its avian jukebox.

A Greek nymph known for her chaste ways gave the organ at the heart of birdsong its name. Syrinx was bathing along a river when the god Pan eyed her and made his move. As she fled, Syrinx encountered other river nymphs whom she begged for help to escape Pan's embrace. Just as the lusty god reached her, the nymph was magically transformed into a tall stand of hollow reeds that played a haunting melody as Pan's frustrated breath blew across their tops. Entranced with the tune, Pan cut the reeds down and fashioned them into his signature Pan pipes.

A bird's syrinx is a very small but complex framework of bone and cartilage that supports the tissue that creates song. It works something like the reed in a clarinet. As the bird's breath passes through it, the tiny membranes in the syrinx wall vibrate to create the song, and the bird adjusts the tension to change its tune. The more elaborate a bird's array of syrinx muscles, the more complex songs it is able to produce. Birds keep singing by taking small and continuous breaths. By changing the position of its neck, throat, tongue, and beak, a bird can change the resonance of its tune.

The human equivalent of the syrinx is the voice box, or larynx. The larynx, though, is found at the top of the trachea, while the syrinx is at the bottom of the bird's trachea, just before it splits in two above the lungs. This configuration allows birds to pass air from each lung lobe separately through the syrinx to create its extraordinarily complex melodies. Songbirds have the fastest-operating muscles in the animal kingdom. The muscles that control their syrinx operate a hundred times faster than the time it takes a human eye to blink.

Singing is mostly, though not entirely, the province of male birds, and it serves at least two critical purposes: It's a signal to females about the reproductive vigor of the crooner—a healthy song indicates a healthy partner—as well as a warning to other males that this territory is defended. The territorial call is believed to be a proxy for bird battles, a way to conserve the energy that would be expended in a physical fight. As for the love song, after a few days of careful listening, female birds make their choice, hearing something about a male's fitness in his voice. Perhaps it's similar to the way some women swoon when they listen to Frank Sinatra or Robert Plant or Justin Bieber sing. And yes, size does matter. The size of the repertoire, that is. Females are far more willing to copulate with males that have larger song repertoires.

One of the still unanswered questions about birdsong is *why* birds sing. Is it only about sex and territory, or are there other reasons? For pleasure, perhaps? Or do birds sing for reasons that lie beyond our understanding? The mythical thorn bird, as the Celts told it, never sang a note; it simply searched its entire life for the sharpest thorn. When it found that perfect thorn, it impaled itself on it, singing, as it died, the sweetest song ever heard. Ofer Tchernichovski, who studies vocal learning in birds at Hunter College in New York, believes they sing for more than practical reasons. He tells a story about seeing a robin on a subway platform that appeared to be ill, possibly dying. Nonetheless, the nearly motionless bird was focused intently on singing what might have been its last song. "It was kind of touching," the scientist says. "He was definitely focused on singing, even though the song was not directed at anyone. It's as if the bird was somehow comforting itself. It seemed to be more encouraged, even in sickness, by its singing."

Taken as a whole, the world's collection of birdsong is an astounding natural symphony, a global treasure of tunes, each one the product of millions of years of evolution, each telling a complex story about that bird and its relationship to others and to its home. And scientists have only begun to unpack those stories. Unfortunately, birdsongs are changing and disappearing as the natural world is altered and destroyed and the human dimension grows noisier and more invasive. That's why the Macaulay Library serves as a kind of Noah's ark, keeping copies of its recorded birdsong collection stashed in a limestone cave for safekeeping.

The absolute peak of the phenomenon of birdsong is the dawn chorus. Just as the sun is starting to rise, especially during mating season in the spring, songbirds across the planet create the world's premier aural spectacle. Because the early morning air is usually still, a song broadcast at dawn is some twenty times more effective as a way to communicate than at other times of the day. One can imagine the exuberant ripple of trills that erupts as the leading edge of sunrise slowly moves across the planet. "Nature's daily miracle," as it's known, is even a holiday—May 5 is International Dawn Chorus Day, when people rise in the darkness and head out on treks to be present at the bird world's first musical stirrings.

There are superstars among the world's singing birds. The hermit thrush sings forty-five to a hundred different notes, with fifty changes in pitch. The aptly named superb lyrebird, an Australian ground nester with an elaborate plumed tail, has the most sophisticated song. Its syrinx features are the most elaborate, which make its tune inordinately complex and give it what is believed to be the loudest birdcall. Moreover, along with seven elements of its own song, it mixes in samples of other birds' songs, along with mimicry of such things as koala grunts or screeches, and even human-created sounds such as sirens, crying babies, and, ironically, the radios and chain-saws of loggers who have come to destroy their habitat.

The sedge warbler, which migrates among Africa, Europe, and Asia, has fifty or so separate pieces to its chattering song, including snippets of mimicry from more than seventy other bird species. A male may never repeat the same song during the course of its life.

The large nocturnal kakapo of New Zealand—whose name is Maori for "parrot of the night"—sucks in air with a shrill whistling sound to fill its two football-size lung sacs, which it then releases in a loud, low-frequency boom. Its song broadcasts more than three miles, a feat the kakapo accomplishes with a natural amplifier: an amphitheater-like bowl that it scoops out of the earth and then positions itself in for maximum acoustic effect. The males may do this every night for months to lure a partner. Unfortunately, night parrots are in steep decline. Once upon a time, kakapo were found everywhere in New Zealand. An early explorer described them as surrounding his camp, "screaming and yelling like a lot of demons" in the night. Now just 126 of these birds remain, and they are under intense management to keep them extant.

Other superstars of the bird-singing world are the male and female plain-tailed wrens, which live high in the Andes, where they flit among the slender green stalks of bamboo forests and sing a unique cooperative duet. The pair take turns rapidly producing notes, three to six per second, and they sing so perfectly in synchrony that the song sounds as if it's coming from a single bird.

It's not only singing that intrigues researchers. Chickadees ostensibly invented social media a long time ago, their very own kind of Twitter, compact bursts of information shared back and forth. Chickadee language—both calls and song—is the most sophisticated animal language in the world, and scientists have spent decades studying it. Cracking the little bird's code will not only tell us more about what they are yammering on about as they hop along the sidewalk and through the bushes; it could also tell us a great deal about ourselves. The chickadee "is a window into the evolution of our own language and our society," says Jeff Lucas, of Purdue University, who has spent decades deciphering the bird's calls.

The black-capped chickadee is ubiquitous in the northern half of North America. Like most birds, the family Paridae, which includes chickadees, tits, and titmice, have two songs, one that expresses territoriality and another that seduces. It's the separate chickadee call system, however, that is unlike that of any other species, and it arises from their unique social structure. Chickadees form pairs in the spring, though once they finish mating they gather to hang out in flocks of anywhere from two to fifty birds. As this tiny tribe hops through the brush, they chitchat with their flockmates, possibly telling of food they found, saying "I am over here," or sounding the alarm if an owl is spotted.

Chickadees can be remarkably specific. The more fearsome a predator to the chickadees, for example, the more *dee*s at the end of their alarm cry. The pygmy owl, a small, fierce, and agile predator that loves to gobble up the birds, causes them to add up to twenty-three extra *dee*s on the end of their cry. The great gray owl, a bird that is far larger than a pygmy owl but which seldom eats chickadees and thus is less of a threat, earned only an extra half a *dee*. Alarm calls, then, are not about the size of the bird posing the threat but about the size of the threat. Such specificity was once thought to be solely the province of human language. "Their social system selects for an immense amount of information transfer," says Lucas, "which in turn selects for an extraordinarily complex call system."

The Carolina chickadees Lucas studies have six note types. The extraordinary part is that they consistently combine and recombine these notes, similar to the way humans organize phonemes to make new words and sentences. That means that the birds understand syntax, and that they have rules of chickadee grammar for their sentences. The calls vary greatly, Lucas says, "from really short calls to ridiculously long calls with up to fifty notes in them." Moreover, the possible recombination of notes and the number of calls they make to each other seem limitless, which is a clue to its sophistication—open-endedness being a defining feature of human language.

Chickadee lingo seems to be nearly universal, a kind of Esperanto of the animal world. "If you record a chickadee here in Montana calling out about a predator and play it in Japan, the birds will all go like this," says Erick Greene, a biology professor who researches bird communication at the University of Montana, cowering and looking up nervously. "And it's not just birds that respond," he says, but "squirrels and chipmunks as well."

Beyond the obvious things—food, predators, seduction, and the like—what else might a chickadee be going on about? The weather? The role of the chickadee in the universal order? "In thirty years I bet we're having a completely different conversation," says Lucas, "because we are slowly getting a feeling for how amazing these creatures are." Nowicki, though, has his doubts about the loftiness of chickadee chat. "Personally, I don't think they are reciting Shakespeare," he says.

While we can't converse with birds in their lingua franca yet, birds can be taught to speak with people. The most famous bird communicator was Alex, an African gray parrot who was owned by a researcher named Irene Pepperberg, a comparative psychologist at Brandeis and Harvard. Alex was unusually brilliant, even for a gray parrot. (Parrots are another bird that researchers think may have a mind.) Alex's speaking abilities revolutionized the way people regarded bird intelligence. He didn't just "parrot" what people said, he possessed a vocabulary of more than a hundred words, which he could put into the appropriate categories and which he could use to express himself. He even made up his own words, for example, giving an apple the name "banerry"—a combination of two other fruits he knew, banana and cherry. Alex died on September 6, 2007, at the ripe old age of thirty-one. His last words to Pepperberg as she left the lab that night, as always, were "You be good. See you tomorrow. I love you."

Linguists have also been helped by the chattering bird class. One current notion is that while early humans used a grunt language similar to that of apes, they blended those grunts together in a structure very similar to the phrasing of birdsong, which created a milestone in evolution: human speech. Our speech has two parts, a lexical layer, or the meaning of a sentence, and an expression layer. Take the sentence "Tom ate an apple." The meaning of the individual words is always the same. The expression layer arranges and rearranges the grunts in different ways to say different things, in the same way birds rearrange their notes to sing different songs with different meanings. "Tom ate an apple!" has nearly the same words as "Is Tom eating an apple?" but in a different order and different expression, giving each sentence a different meaning.

Something may be hardwired in us to hear and be deeply moved at fundamental levels by the ubiquitous music of birds. I am moved when I hear the trill of a meadowlark in the meadow near my house. In birdsong, as in birds' flight, I experience some of the lightness of being we so crave. The romantic poet John Keats wrote of his passionate love for birdsong in "Ode to a Nightingale," seeing it as nothing less than an expression of immortality—"pouring forth thy soul abroad in such ecstasy!" he wrote. Percy Bysshe Shelley in "To a Skylark" waxes rhapsodic about birdsong as an emanation from heaven, a balm to escape the pain of life:

Teach us, Sprite or Bird,
What sweet thoughts are thine;
I have never heard
Praise of love or wine
That panted forth a flood of rapture so divine.

Birdsong has inspired a wide range of music, from Boccherini's *The Aviary* to Mussorgsky's "Ballet of the Chicks in Their Shells" to Tchaikovsky's *Swan Lake* and Vivaldi's *Spring*. Mozart borrowed notes from the song of a starling he kept, and when it died he was so broken up that he threw the bird an elaborate funeral. And the opening notes of Beethoven's Fifth Symphony are remarkably similar to the song of the white-breasted wood wren.

The most devoted of the birdsong composers was Olivier Messiaen, the mid-twentieth-century avant-garde musician for whom birds were the pinnacle of musicality and an expression of the spiritual. "They are our desire for light, for stars, for rainbows, and for jubilant songs," he said. Messiaen had a nervous system anomaly called synesthesia that, in his case, was bidirectional—that is, he "saw" music in his mind represented as beautiful swaths of color, and he "heard" color as music. A walk through the woods watching and listening to the birds, then, must have been quite a treat for the Frenchman.

Messiaen was as much an ornithologist as a composer, and he traveled the world listening to a wide range of exotic birds and then writing pieces based on their songs. He didn't just replicate their music, he also worked the singing into complex and atmospheric tone poems. He quietly strolled through the woods in the French Alps where he lived, or through the exotic bird markets of Paris, scribbling the notes of hundreds of different birds of the same species, to create a composite and ideal warbler, for example. His first birdsong piece was *Oiseaux Exotiques* (Exotic Birds), and it is considered a landmark piece for its precise use of birdsong. Another composition for the flute, called *Le Merle Noir,* was based entirely on the songs of the blackbird, which Messiaen felt had the sweetest tune of all.

Why does birdsong affect us so? The male's song is shown to increase dopamine, a pleasure chemical, in female birds; perhaps it triggers a cascade of reward neurochemicals in us as well.

A number of businesses sell the sound of birds for its purported

psychological benefit and ability to boost productivity. Julian Treasure runs the Sound Agency, selling companies on the idea that a
birdcall is part of our deep past and so has a profound impact on
productivity, encouraging a state called "body relaxed, mind alert."
"People find birdsong relaxing and reassuring because over thousands of years they have learned when the birds sing, they are safe,"
Treasure told Denise Winterman of BBC News. "It's when birds stop
singing that people need to worry," because their silence may mean
danger is nearby. "Birdsong is also nature's alarm clock, with the
dawn chorus signaling the start of the day, so it stimulates us cognitively."

A persistent notion runs through some of the world's esoteric traditions that holds that in the distant past, birdsong was a universal
language understood by humans and birds alike. Precisely what this
language might have been is hard to tell, for there isn't much to go
on, just a smattering of oblique clues in several different texts. In
Sufism's mystical poem *Conference of the Birds,* Attar the Chemist
describes such a language, and in the Talmud, Solomon's wisdom is
said to come from the birds. In the collection of anonymous Norse
poetry known as the Poetic Edda, Sigurd, a legendary hero of Norse
mythology, tastes dragon blood, and it allows him to understand
bird language. In the Kabbalah it is known as the secret and perfect
language.

Mythology holds that Hugin and Munin, the ravens who sat on
the shoulders of the Norse god Odin and were his eyes and ears in
the world, reported to him in this language. Birds, some say, used
this language to talk to humans initiated in spiritual matters and
vice versa. Shamans are said to speak it in their trance states to communicate with birds and other animals. According to a paper published in the *Journal of the Western Mystery Tradition* in 2003 by
the scholar Vincent Bridges, Jesus and other ancient mystics were
purported to have had command of this secret bird language. Saint
Francis of Assisi, the twelfth-century Catholic friar who believed nature mirrored the divine, was conversant in the language of animals,

preaching often to "his sisters, the birds," who were said to be enchanted by the sound of his voice.

Cyrano de Bergerac wrote about a traveler who encountered a brightly colored bird wearing a golden crown. The bird sang to the traveler, who was surprised to find he could understand bird language. "There are to be found among the birds those who can speak and understand your own language," the bird in Bergerac's tale sings. "Thus, just as you will encounter birds that do not say a word, others that merely twitter and others that can speak, so you may even encounter one of the most perfect birds of all—those who use all idioms."

According to Bergerac, the cathedral-building Freemasons were among those who spoke in the rhythmic language of the birds and camouflaged the esoteric language within their everyday conversation so that it was clear only to those who understood it. "What unsuspected marvels we should find," he wrote, "if we knew how to dissect words, to strip them of their bark and liberate their spirit, the divine line, which is within."

Bergerac's belief echoes the theory of scholars of the esoteric such as Vincent Bridges and the French alchemist and author Fulcanelli, who believe bird language is a phonetic kabbalah whose sounds are alchemical, a divine spark that somehow activates human DNA to evoke and maintain a spiritual awareness in which the language of the birds can then be understood. "The language of the birds," writes Fulcanelli, "is the common language of initiation and illumination behind cultural expressions as different as the Christian, the Inca, the medieval troubadours and ancient Greeks. And traces of it can be found in the dialects of Picardy and Provence, and most important of all, in the language of the Gypsies."

Climate change looms as the biggest threat to birdsong. In a grim, illustrative 2012 piece called *When Birds Sing a Toxic Sky,* the London performance artist Liam Young placed eighty birds in a sealed room in Amsterdam. Wearing a gas mask, he opened the valve on a tank of carbon dioxide and slowly raised the level inside the room from 360 parts per million, around where we were then, up to 1,000 ppm, which is the level the earth is predicted to reach by the end of the century. At first the birds sang merrily away, but as the levels rose, the singing slowed, rhythms were altered, and finally the singing stopped entirely. The birds didn't die, they simply stopped singing.

Do birds have more to tell us about the world and ourselves than we know? Are we simply unable to comprehend their vital messages because we have lost the knowledge? Can we regain it? The answer seems to be yes—if we keep them around to find out.

# CHAPTER 13

The Bee-eaters: A Modern Family

> How helpless we are, like netted birds, when we are
> caught by desire.
>
> — BELVA PLAIN

If you wanted to travel far back along the human timeline and understand how, over many thousands of years, early humans came together to bond in harsh, unpredictable environments to mate and have children and form bands of extended families—grandparents, uncles, aunts, and siblings—who all worked together to raise the kids to assure their survival, where would you turn? If you wanted to shed light on the dynamics of stepfamilies—why some stepparents aren't as kind to stepchildren as they are to their own, or why stepdaughters are more frequently sexually abused by stepfathers than biological children—where would you look? You can't study how evolution drove these behaviors among people because culture and biology are inextricably tangled. So where do you look for a pure biological signal? And what kind of scientist would best search out the origin story of families?

An ornithologist might not be the first to come to mind. And a tribe of rainbow-hued birds known as white-fronted bee-eaters, who live in holes they dig in sandy cliffs on an arid landscape in Lake Nakuru National Park in Kenya, might not be the first critters to spring to mind as a proxy to study human families. These birds, though, and the tangled web of love, lust, deceit, child rearing, and adultery that makes up their lives, are exactly what Stephen Emlen, a distinguished ornithologist and evolutionary biologist at Cornell University, has long used as a model system to penetrate one of the deepest and most important subjects of evolution. Emlen believes that the way bee-eaters came together to cooperate as family groups to survive in a marginal environment is very similar to the way early human families banded together.

There is, it seems, a little white-fronted bee-eater in all of us.

Emlen, now seventy-four, has been researching birds since the 1960s and studied the bee-eaters in the 1970s and 1980s. As a young researcher he designed, with his father, the ornithologist John T. Emlen, something called the Emlen Funnel to study the mysteries of migration. The funnel-shaped plastic device had a piece of paper at the bottom coated with ink. A bird was placed in the bottom of the funnel, which in turn was placed in a planetarium. Then the sky in the planetarium was rotated, and as the constellations changed, so did the orientation of where the bird jumped up to try to exit the funnel, as evidenced by its inky tracks. The experiment demonstrated that migration is governed, at least in part, by celestial alignments.

It was the family life of birds that really hooked him, though, and he has spent most of his life looking at how environment has shaped the evolution of animal and human societies. He researched the synchronized breeding of bank swallows and canoed through the swamps of Panama trying to understand the gender idiosyncrasies of the jacana—a long-legged bird with oversize feet, also known as the "Jesus bird" or "lily trotter" because it walks from lily pad to lily pad across the water. The jacana is a curious bird. Females are the larger and dominant of the sexes, while males are house husbands, tending to the eggs and young. And the ladies practice polyandry, keeping a harem of up to five males, because some of their guys inevitably disappear into the jaws of a crocodile.

It is among the bee-eaters and their soap opera lives, however, that Emlen made his bones as an evolutionary ornithologist. The world has twenty species of these scrappy little insect hunters, mostly in Africa and Asia, but there is also a smattering in Europe and Australia. They are a tough bird on a tough landscape, known for facing down a dangerous puff adder, a deadly viper that threatens their brood. Their name comes from their predilection for gobbling up the stinging bee, which they catch on the wing, though they eat other kinds of insects, too.

The white-fronted bee-eaters of Kenya, like the other bee-eaters, are visually striking birds, similar in shape and size to the kingfisher, another member of the Coraciiform family. Male and female bee-eaters have a similar appearance—a green back and cinnamon or yellow bottom, a black mask, a white forehead, a band of blue around their bottom, and a bright red throat that looks as if it were airbrushed on. They have a long, sharp, slightly downturned bill, which operates like a pair of forceps. When they snag a wasp or bee, they alight on a perch and whack it against a branch to kill it. Then, eyes closed, the bird rubs the insect vigorously on the bark to expel the poisonous venom before downing it.

Bee-eaters live in communal quarters that they create by loosening dirt in a sandy cliff or in the ground with their jackhammer beaks and excavating the loosened soil with their feet. Then they burrow in, creating nesting holes up to five feet deep. As many as fifteen to twenty families—made up of some one hundred and fifty individuals in total—live together in one community. While the birds range widely across the savanna to feed on bees and other insects during the day, they return here in the afternoon to socialize and roost.

The most unusual thing about bee-eaters is the way kin and friends swarm around breeding parents, eager to lend a hand. While about half or more of the birds breed, the rest are ad hoc nannies and play a major role in every aspect of birthing and rearing offspring except copulation. Assistants deliver food to the females while they are laying eggs, and some helpers even undergo physiological changes that allow them to incubate eggs. When the mother leaves the nest, the other family members pitch in as fill-in helicopter parents, defending the youngsters as they face threats at home or when they wander away from the nest. And the more harsh the environmental conditions, the higher the percentage of birds that do not breed and instead help other pairs.

The intriguing family dynamics of the bird first attracted Emlen in 1973. He had heard that the bee-eaters were unusual because they were a rarity in nature—an altruistic society. A cardinal rule in evolution is that of the selfish individual, who spreads his DNA at all costs so that his genes survive. Could there be a bird that sacrificed this evolutionary imperative for the greater good? Emlen spent a decade in the wilds of Kenya studying the hundreds of birds that make up the extended families of the bee-eaters. He still lectures about lessons learned from them.

When Emlen's work on the project first began, he and his wife and colleague, Natalie Demong, trapped the birds at night by placing mist nets over their holes so they could mark them with a small tag. That meant walking barefoot through the grass to avoid crushing eggs—despite the presence of those pesky and poisonous puff adders. Stories like that, and his paddling among the crocodile-infested swamps of Panama, earned Emlen a reputation as a swashbuckling bird nerd—an image he downplays. If you love doing something, he says, those kinds of annoyances don't matter. And he is passionate about his work, which he likens to the role of a detective, one who monitors the birds intimately over a large portion of their lives: "You develop ideas of what might be going on while watching, and it takes years to collect the data that could potentially refute or support those ideas. It's very exciting to be an ornithological detective."

The researchers participating in Emlen's study set up their shop a few hundred yards from the bee-eater village. They felt, he told me, a bit like Jimmy Stewart in the Alfred Hitchcock film *Rear Window,* in which an injured and house-bound race car driver watches his neighbors' lives through their windows. The researchers meticulously photographed and recorded the birds' homes, so that each bird family had a known "address." They crafted a special tool out of dental mirrors and a light that they could stick into the holes to see what was going on inside. They noted the comings and goings of the whole family, as well as sundry other details and behaviors: who had food and who didn't, who greeted whom. They recorded courtships, births, and mortality. Nests with eggs got special attention, and they watched the doting parents-to-be in the nest with binoculars for hours on end, and again when the hatchlings were young, and a third time when the birds were ready to fledge. They gathered blood samples and developed bird genealogies based on DNA assays, and with those they created a complex family tree of relationships for the bee-eaters of Nakuru.

Over the decade of their field research, the team found several core bee-eater family principles. First, the birds form large multigenerational families, from three to seventeen members of genetically related kin. A typical bee-eater family has two or three reproductive pairs, plus several birds without mates. Along with parents and young, the family group includes siblings, grandparents, uncles, cousins, and often even steprelatives, because birds mate again after a spouse dies. They even go through divorce. How the birds in these families interact with one another and how living among extended family shapes the individual are the questions that lie at the heart of Emlen's work on the evolution of extended families.

Familial groups are a rarity in the animal kingdom, but they are most common among birds. About 10 to 12 percent of bird species worldwide, some 850, practice cooperative breeding—which means three or more individual birds rear young together. The vast majority are nuclear families, which means they are comprised of a father, mother, and one, two, or several nonbreeding children. Families of extended kin living together, like the bee-eaters, are very rare. Only some twenty species of birds are known to live this way.

A major difference between a nuclear and an extended family in the bird world is that in an extended family, a "kid" will grow up, take a mate, and reproduce, yet continue to live at home with its parents, grandparents, siblings, and other kin. By helping to raise one another's babies, the extended family group is able to be successful even when conditions are harsh and food scarce. By working together and helping to raise the babies until the rains come, the group can provide enough food to get through difficult times. When rain does arrive, causing plant growth and bringing out worms and insects, it provides enough food to allow the kids to strike out on their own. The young birds in a nuclear family, on the other hand, may help rear younger siblings but don't mate or reproduce until they leave their parents.

Emlen is both an ornithologist and an evolutionary biologist, which means he is concerned about how the environment in which creatures evolve shapes their survival, reproduction, and behavior. The extended bee-eater families of sub-Saharan Africa, he says, may be just like the first human families that evolved in the same part of the world. "If it was a godawful year, these human extended families could link together and help one another," Emlen says. "If it was a good year, the family could increase in size by having several different breeding units all reproducing." In other words, the size and cohesiveness of the human family was affected by changes in rain, food, and other environmental factors in the same way the populations of bee-eaters are affected.

Some scientists think the emergence of cooperative breeding explains the development of our emotions and high-level cognitive skills, which led to a social intelligence that enabled nothing less than modern society to form. Sarah Hrdy, a professor emeritus of anthropology at the University of California, Davis, was one of the top researchers of cooperative behavior in primates. She believes that the evolution of the seemingly irrational "prosocial" behavior we engage in to get along—irrational because it's not, as the theory of evolution demands, selfish, such as helping mothers feed their children, forgoing territoriality, babysitting the young ones while the parents look for food, and otherwise cooperating for no apparent benefit—allowed *Homo sapiens* to thrive, spread out from Africa, and conquer the world.

"One of the steps along the way to higher cognitive abilities is to recognize and do things that lead to the common good," Emlen says. "These birds know a large number of specific individuals, because they tolerate them and roost together at night. But they defend every nest entrance and don't let in others who are not part of their extended families. Bee-eaters seem to have higher and more complex cognitive functions, and that is due to the complex society they live in." Moreover, knowing and caring what others think of you and how they will respond to you, being conscious that other creatures have an awareness and feelings similar to yours, is the basis for the Theory of Mind, the scientific idea of what it takes to be conscious. It's at the core of our civilized society.

At first Emlen thought he had discovered a bird Shangri-La, entirely free from selfishness. Birds that don't have babies as soon as they are able, that stay home and live in the avian equivalent of their parents' basement and help rear their brothers and sisters, would seem to defy the idea of a competitive reproductive strategy to perpetuate their DNA. How did genes that weren't that interested in reproduction get this far?

He found, however, that the birds are not altruistic, at least not in the way we think of it. Theirs is simply a different strategy, a cooperative one to assure fitness instead of a competitive one—a selfish kind of altruism that helps perpetuate the genetic material they all share. Having just one helper, Emlen's studies show, doubles the number of successful fledglings.

But this tribe of bee-eaters is a long way from an idyllic *Leave It to Beaver* land. Over the decade of watching, Emlen learned that, to a remarkable degree, the communal bee-eaters exhibited many of the behaviors that take place in the human carnival of life, from courting to bullying to sly forms of deceit. Fooling around—or, as the ornithologists like to say, extra-pair copulation—does occur among the bee-eaters. Male bee-eaters even forcibly copulate with female birds, something that occurs in other bird species, too. "It is kind of brutal when you see it," Emlen says.

There is also the all-too-human behavior of not wanting the kids to leave the nest—though here it's about maintaining the extended family for reasons of survival. When young birds get that look of love in their eyes, the parents will suddenly become very attentive, to the point of distraction—which is entirely the point. A relative might show up to visit often to disrupt the young bird's feeding of his mate or guard the entrance to the nest burrow and not allow the female bird to pass. It's a way of stopping the union so the youngster will return home and again assume a role as a helper.

Bluebirds also hold a mirror up to our society's mating habits. While monogamy has long been held up as an ideal in most of human society, it was once upon a time thought that many species of birds also faithfully paired. But in the 1980s, Patricia Adair Gowaty, now a professor of evolutionary biology at UCLA (and a former postdoctoral student in Emlen's lab), did genetic testing on bluebird offspring—the first of its kind. Lo and behold, among one of the paragons of pair-bonding, she discovered that 15 to 20 percent of the bluebird chicks cared for by a mother and father were really fathered by a different male. This was deeply shocking to ornithologists when it was first discovered, and for years, Gowaty says, many researchers denied her work, even accusing her of making up data, because it upset the notion of faithful bluebird pairs. Now, however, it's widely accepted that bluebirds—and many other species—spread the love around more often than not. "Genetic mo-nogamy is the rarest breeding system in birds," she says. "Females mess around."

One of the reasons no one suspected the bluebird is that the wham-bam nature of passerine sex meant their extra-pair copulat-ing went unnoticed. Instead of the penis-vagina model, they squeeze a small sac out of their cloaca—which is used for excretory function as well—and touch their pouches together, something politely known as a cloacal kiss. "It takes about four seconds," says Gowaty, "and is very easy to miss."

These tendencies to fool around might explain some human be-havior, she says. While monogamous marriage among adults is a social contract, our biology, a product of evolution, has a different agenda. "Evolution and human society are in conflict," she says. "There are time-honored ways to have healthy progeny, and one of them is called the lottery mechanism. That means that you don't buy ten lottery tickets with the same number, you diversify so you have a better chance of winning. Extra-pair copulation, then, might give an evolutionary advantage." A good excuse for extramarital sex might be "Evolution made me do it," she says.

Bluebird relationships also teach us that dalliances soar when nests are closer together. In nests nine meters apart, extra-pair pater-nity occurred at a rate of 35 percent, while in nests a kilometer apart, it was 8 percent. So it's a function of opportunity, a conclusion, Gowaty says, that "is exportable to humans."

It was while watching bee-eater behaviors play out that Emlen realized that if humans understood that many of these behaviors are welded into us by the blowtorch of evolution, encoded in our genet-ics, we would understand a great deal more about who we are, why we make the decisions we make and behave the way we do, and how we might head these behaviors off if they work against our best in-terests. That's why Emlen spends much of his time translating the lessons of the bee-eaters to the human condition. He feels so strongly about the parallels and the potential we have for learning from them that he took off his ornithologist's hat to present a "unified family theory" in a paper called "15 Predictions of Living Within Family Groups," which he believes should apply to all families, whether they are insects, birds, or mammals, including humans. He also joined with colleagues to form the Evolving Family Project at Cornell University, which held conferences bringing together practitioners of the evolutionary and social sciences.

Some of his observations have been controversial. In fact, four of the fifteen bee-eater-inspired evolutionary principles that Emlen lays out are about stepfamilies, and these have gotten the most attention and engendered the most controversy. The first is that stepparents are less invested in their stepchildren than they are in their biological children—the so-called Cinderella effect. Because only biological children carry their birth parents' genes, there is an unconscious tendency to favor those children to make sure they survive. "You can't just bring in a surrogate parent and expect that a two-adult family is going to have the same effect as a two-biological-parent family," he told me. To have a family that functions as highly as possible, it's best to have two parents tightly and biologically bonded to their children. And research since has overwhelmingly supported that notion, he adds. Also, step- and half siblings will not get along as well as biological brothers and sisters because there is no shared genetic inheritance to protect. Perhaps the most controversial finding is that bee-eater stepparents are more likely to have sexual relations with their stepchildren. In bird families related by blood, incest is extremely rare because it exacts a "fitness cost"—that is, it causes inbreeding, and so a lower rate of survival, in the chicks. Not so in stepfamilies—human or bird. Emlen cites data that show that sexual abuse of human stepdaughters by stepfathers is eight times that in biological families, and he believes it is driven in part by the unconscious knowledge that there is no fitness cost. When he revealed his findings, "I got hate mail from people that said things like 'I grew up with a stepfather and I wasn't abused,'" Emlen says. "And that's not what I am saying. I am saying abuse is predicted to happen more often, but the vast, vast, vast majority of stepfathers form close bonds with their stepchildren and do fine."

And it's not just stepfather and stepdaughter who tend toward incest. In bee-eater society, should a male bird's mother die and his dad take a new partner, the son will try to mate with his stepmother at every opportunity. Should the father die, however, the son will turn on his new mother, often banishing her, in order to assume a leadership role. And, because of these factors, stepfamilies—human and bird alike—are inherently less stable, which means higher divorce rates and higher probabilities of behavioral, emotional, and health problems.

Other lessons from the bee-eaters? One takeaway, Emlen says, is that the idealized human nuclear family—mom and dad and two kids—is an evolutionary anomaly, "a culturally novel type of family." And single-parent families, he says, are even more of an "evolutionary oxymoron." They simply don't exist in the bird world. Large extended families, and the care they provide—"a built-in workforce" that includes grandmothers, grandfathers, uncles, and aunts laboring to protect the family genetics—are the evolutionary norm and the kind of family structure human beings have lived in for thousands of generations. During most of human history, if one parent died or otherwise left the group, Emlen says, extended families remained to help care for the young.

Emlen's ideas have been criticized for implying that dysfunction is genetically determined rather than a character flaw, that it gives people an excuse for misbehaving. That is a misinterpretation, Emlen says. Evolutionary tendencies are just that: a tendency or a predisposition, not a done deal. A main benefit of understanding the behavior of bee-eaters, of viewing them in their pure state in nature, without cultural overlays, is that if the same behaviors are recognized among people by therapists and their clients, they can be mindfully overcome. He likens it to the revolution in genetic testing— knowing you are predisposed to a disease allows you to take preventive measures. "A Darwinian perspective may train people to be on the alert, to have their antennae out," he says. "If one can screen individuals and tell them they're at an increased risk for a particular disease, they may change their eating habits or their behaviors to lower that risk." That logic can be used in the counseling of stepfamilies. An awareness that stepparents have more difficulty forging the same emotional bond with a stepchild as with a biological child can help a family to understand and overcome the issue.

Emlen has always been up against the skepticism of people who can't accept that birds have anything at all to teach us about human family dynamics. Primates, he says, people get, but not birds. "It's as if there's a cement wall that people can't cross," he told *Discover* magazine. "And I see my role in breaking that wall."

Martin Daly, a professor of evolutionary psychology and anthropology at McMaster University in Ontario, who has written a great deal about stepfamilies, including *The Truth About Cinderella: A Darwinian View of Parental Love,* has incorporated some of Emlen's research into his thinking and writing. "Anthropologists don't pay as much attention to bird behavior as they should," he told me. "The only nonhuman animals that they seem to consider 'relevant' are other primates. That's unfortunate, because birds are often a better source of insight into human predicaments."

"I couldn't have agreed with him more, that bird relationships resemble human relationships and family systems more than primates do," Michael Kerr, who runs the Bowen Center for the Study of Family at Georgetown University, told me. The center was named after Murray Bowen, a research psychologist at the National Institutes of Health who developed the Bowen Family System Theory. One of its key concepts is a triangle representing interpersonal dynamics, such as when two people in a family have differences and one of them seeks support from a third. A mother who is at loggerheads with a father, for example, may come to rely on a child for support. Emlen found examples of this in bee-eater society. "The mother of a bee-eater offspring had died," Kerr said, "and the father was without a mate, and one of the male offspring was attracted to a female from another family. The father was actively interfering and stopping the effort of the son to relate to the female. It's a kind of triangle, and we find it in people as well."

The positive aspects of being part of a large and nurturing bee-eater family led Emlen to become an advocate for the restoration of traditional extended human family relationships. "Society is trying to cope in many ways with the absence of this built-in set of relatives who are prewired to help," he says. He likes the Australian idea, for instance, of tax breaks for "granny flats," a term for apartments where grandmothers can live to be near families to help care for children. "I got in trouble on that one, too," he says with a laugh. "People wrote and said things like, 'I have no interest in living near my grandmother.'" Emlen has also called for more funding for research on the role of grandparents in the family.

Meanwhile, a different species of bird has shown that wealth management is a glue that keeps families together, evolutionarily speaking. The acorn woodpecker, with its black-and-white body topped off with a crimson skullcap and a pair of wide yellow clown eyes, lives in large family groups in oak forests in the western United States. Acorn woodpeckers store acorns in small cavities they have drilled into trees, pounded into fence posts, and carved out in other wooden objects, and sometimes they make hundreds or even thousands of holes. The bird carefully inserts an acorn into a hole that matches its size, and after the acorns dry and shrink, they are moved to smaller holes. When times are good, these birds expand their caches into other trees, building an acorn empire that can last for generations; and if the living is easy, with many trees filled with many acorns, then a young woodpecker often delays his departure from home, similar to the way human families work together over generations to preserve their wealth. Such bird—and human—families are known as "dynastic."

The notion of cooperation among birds, which leads to the greater good of family and society, has inspired some evolutionary biologists to see another lesson: that altruism is sometimes how nature works, even if it is altruism for selfish reasons. These birds, they say, might point the way toward a better human society. "Look carefully at nature and you will find that it doesn't always seem short, brutish, and savage," said Helena Cronin, a philosopher and social scientist at the London School of Economics who studies extended animal families and the things they have to teach us. She was addressing an elite crowd at the World Economic Forum in Davos, Switzerland. "Animals are strikingly unselfish, giving warnings of predators, sharing food, grooming one another, adopting orphans, fighting without killing—or injuring—their adversaries. In some ways they behave more like moral paragons of Aesop than the hard-bitten, self-seeking individualists that natural selection seems to favor.

"What if being the fittest means having the most generosity of spirit?" she asked. "What if enhancing your chance of survival comes from improving your capacity to be altruistic? The more people understand that we are evolved altruists, and the more people feel that no one is taking advantage of another, the more we will become altruistic, and the more we won't take advantage of one another."

How will the very rare birds that raise their young cooperatively fare in a changing world? For the extended families of bee-eaters and social weaver birds that Emlen studies, it could fall one of a few ways. Severe changes in climate where bee-eaters live could decimate populations. Or weather could improve in those marginal parts of the world, and more rain, for example, could provide more food, which would help the birds thrive. Or, Emlen believes, the extended family model could be resilient and adaptable enough that the birds may be able to withstand an even harsher world.

But the situation among one set of cooperative breeders, the Florida scrub jays, is more dire. The subject of the longest-running bird study in the country, by John Fitzpatrick at the Cornell Lab of Ornithology and others, Florida scrub jays are threatened by habitat destruction from development in the sand dunes where they make their home. The territory they need for survival is shrinking, and the little that remains is so overpopulated that young birds have no place in which to set up a family. When a paired male dies and space does open up, three or four competing couples will zoom in to try and claim the spot, like too many campers at a campground. "The choices for a young scrub jay are very bleak," says Emlen.

That is a tragedy, according to Emlen, for bird families are the window through which we can look back clearly on the evolution of a wide range of human behaviors, perhaps more clearly than with any other animal. They may have answers to the source of some of our thorniest human problems. They may even be able to weigh in on some existential questions, such as the factors that lie at the roots of crime and war. "Knowing more about how we got here," Emlen says, "is an important step to improving these problems."

# CHAPTER 14

~~~~~~

# Extreme Physiologies:
# Birds, the Ultimate Athletes

Birds know themselves not to be at the center of anything,
but at the margins of everything. The end of the map.

—GREGORY MAGUIRE,
*OUT OF OZ*

According to the New Zealand climber George Lowe, as he strug-
gled through deep snow to reach the summit of Mount Everest in the
1950s he was startled to hear honking in the sky and looked up to
see an extraordinary sight: flocks of bar-headed geese breezing along
toward Central Asia through a crystal-blue sky. This was, mind you,
at more than five miles up, the height of the world's tallest peak.

It's hard to imagine a more punishing and inhospitable environment for birds than at that elevation, with winds blowing at more
than a hundred miles an hour, temperatures plunging to thirty or
forty degrees below zero, and an oxygen level about one-fifth of that

at sea level. The air at that altitude, in fact, is so thin that helicopters cannot fly because their rotors can't get enough lift. Yet the five-pound bar-headed goose, a striking light-gray bird with a white head decorated with two bold black bars, easily glides through the sky here at about fifty miles per hour, and with a stiff tailwind it can travel at twice that speed. Moreover, if the goose catches the right breeze, it takes only about fifty hours to fly from the soggy marshes of India's far south to the mountain lakes of Mongolia, a distance of about a thousand miles. The bar-headed geese cross the tallest mountain range in the world in a single day.

The rate of ascent of these geese is the fastest of any bird, and the rapid change in altitude they make would kill a person. During the highest part of their journey, the geese tolerate an oxygen level only 20 percent of that at sea level, even as they dramatically increase their consumption of oxygen to power their flying. The bar-head owns the title to the world's highest and most extreme migrations. Scientists call it a "superbird."

It's not the bird world's only extreme performer by a long shot. The tough and tiny blackpoll warbler, with its black cap and mottled black-and-white coloring, is another miracle with wings. These birds depart the rugged and storm-whipped shores of New England and the Canadian Maritime Provinces in the fall and head up into the sky until they level out at about a mile up. For the next eighty to ninety hours, depending on which way and how hard the wind is blowing, they push relentlessly toward coastal Venezuela, covering some two thousand miles without stopping. Their exertion and speed are equivalent to a person running four-minute miles for eighty hours, and they are unmatched by any other species.

Then there is the small, slim, and aerodynamic Arctic tern, white with a black mask covering its head and eyes, like Batman, set off by an orange slash of beak. This seabird sees more sunlight than any other creature on the planet. It summers in the Arctic, where it basks and feeds in the twenty-four-hour daylight, and then, as the light begins to diminish, it travels twelve thousand miles south to take in the round-the-clock summer light in Antarctica. And it makes that epic journey without stopping. Across its lifetime of about twenty years, the Arctic tern becomes a platinum frequent flier, journeying more than a million and a half miles.

What specialized physiological qualities has many millions of years of survival-of-the-fittest-driven adaptation equipped these and other birds with that allow them to unerringly find their way across vast and hostile distances in extremes humans cannot approach? That's a question a number of scientists are asking in the hope of learning more about the birds' performance and how those lessons might be applied to people. Migration remains a complex mystery to science, though not for lack of trying.

In 1965, the researcher Friedrich Merkel brought several European robins into his lab for an experiment to answer the great puzzle of how birds migrate. There's a German term, *Zugunruhe,* which translates roughly as "restlessness" and refers to the condition birds display as part of their preparation for migration. They manifest that unsettledness even when caged in a room with no outdoor view. Merkel's experiment was conducted in the fall, when the European robins were beginning their anxious period. Merkel placed a curtain around the birds' cage and placed the cage inside a steel cabinet, to block out the earth's natural magnetism. Next to the birds he placed two Helmholtz coils, lab equipment that creates a uniform magnetic field that was, in this case, identical to the frequency of the earth's field. As *Zugunruhe* intensifies in birds in the wild, they start to orient themselves southward for a longer and longer period each day, preparing for departure. Merkel placed the Helmholtz coil to the east, and the birds, instead of turning south, oriented themselves toward the man-made frequency again and again. Merkel proved that the earth's magnetic field is a main factor in how birds navigate. At first, it was thought he'd solved the riddle. But this discovery, while important, was merely the beginning of an understanding of the complexity of bird migration.

Take the case of European starlings, which depart Sweden each autumn and flap off in a southwesterly direction through Holland, heading for a region of France and Britain near the English Channel. In the 1950s, Dutch biologist Albert Perdeck and his team captured thousands of the birds and banded most of them. They left half of the birds where they had been captured and shipped the other half to Switzerland. The ones left behind headed straightaway for their customary destination near the channel. Among the flock released in Switzerland, the older ones realized they had been hoodwinked and made a correction, ninety degrees from their usual course. The first-year birds, though, didn't correct, and continued southwest. They eventually ended up on the Iberian Peninsula, a long way from their goal. Perdeck believed the birds' usual migratory direction was innate, and passed on genetically, but because only the adults knew they weren't in Holland and needed to fly in a new direction, the journey must have also been somehow modified by experience.

Other studies have taken place in planetariums, including those by Stephen Emlen, which show that birds orient by constellations. And birds wearing tiny GoPro-like helmet cams demonstrate that they key off of natural and human-made landmarks, mountains, roads, and bridges as navigational cues. Iron filaments were recently found in the tissue of birds, and it's believed they may help orient the fliers to the earth's magnetic line. Prevailing winds, odors, and the sun are also factors.

More recently, though, some researchers have come to believe that quantum effects may lie behind birds' ability to make their way across the globe. Quantum phenomena are so weird that Albert Einstein called one type "spooky action at a distance."

So extraordinary is bird vision that birds are often called "a pair of eyes with wings." Birds of prey can see up to eight times better than humans. Owls have particularly sharp vision—and in the dark. Part of the reason for their excellent sight is the fact that many birds have the largest eyes relative to their body size in the animal kingdom, and bigger is better because the image captured on the retina is larger, similar to the way the most richly detailed photo images are made with large-format cameras. Birds also perceive four channels on the light spectrum: the three channels that they share with us—red, green, and blue—as well as one we can't perceive, ultraviolet. Their visual world is ruled by a wide range of ultraviolet colors—in flowers, on insects, and on their mates. We may see a female bird as drab brown, but other birds see striking colors.

A recent idea about how birds may use their vision during migration may be the most phenomenal thing of all about avian perception. It's known as magnetoreception. Researchers at several top institutions have identified a light-sensitive protein called *cryptochrome* in special cells devoted to vision in bird eyes. When particles of light called photons strike the cryptochrome, it charges electrons that fleetingly—for one hundred microseconds or less—exist in a state of what's called quantum entanglement with the earth's magnetic grids. Even though these lines are a fraction of the strength of a refrigerator magnet, entanglement allows the bird to see and be guided by them as it flies.

This possible explanation for way-finding is at the leading edge of the quantum shock wave that is bearing down on the field of biology. Einstein made his remark about "spooky action" because things that were separate—in his case, two distinct atomic particles—were in reality connected, through an invisible "entanglement." Quantum effects defy the laws of physics as we know them. They were once thought to belong solely to the atomic realm, apparent only in precisely controlled laboratory conditions, and not something we needed to concern ourselves with in the larger world. That belief is changing. Quantum biologists are studying how these slippery, difficult-to-fathom principles are manifest in nature around us, and their findings are beginning to show that the prevailing model, which considers only the material realm—those things we can see and measure—fails to consider what lies outside that realm. What we've done with our life sciences to date, it seems, is describe a small portion of a much larger and decidedly stranger world simply because we haven't been able to imagine, let alone study, what may really be going on.

The same light-sensitive protein the researchers identified in birds has been found in the human eye, and some researchers speculate that humans also have the ability to see magnetic lines this way, though our ability is dormant. Understanding how it works in birds may help us learn more about, or perhaps even activate, this unusual power of vision.

Still, despite the wide array of theories, findings, and speculations, no one can yet explain how birds are able to migrate. "Remove cue after cue and yet animals still retain some backup strategy for establishing flight direction," writes Rupert Sheldrake. "The problem of navigation remains essentially unsolved."

Jessica Meir is one of a group of scientists who are teasing out some of the secrets of how bar-headed geese are able to perform so vitally in hypoxic environments, those in which oxygen levels drop to a fraction of normal. Meir, thirty-nine, is no stranger to extremes herself. She served as an aquanaut on the NASA Extreme Environment Mission Operations in Florida in 2002, living for two weeks in an underwater laboratory as a simulation of life in space. She also coordinated experiments to study the physiology of astronauts on space shuttle missions. And not long after an interview for this book she was accepted as a NASA astronaut candidate—her dream job—and moved from Harvard to Houston for training.

Meir's work to understand extreme physiology has focused on both the high-flying bar-headed goose and the emperor penguin, which, when it feeds, plunges many hundreds of feet down into the frigid waters off of Antarctica to forage. It's the deepest-diving bird of all, and it has evolved many exquisite adaptations to feel right at home in ice-cold, extremely low-oxygen environments.

Meir's bar-head work, which started in 2009, looks at how the bird manages to get so much oxygen out of thin air during its flight over the Himalayas. "It's especially impressive, because flying is the costliest form of vertebrate locomotion," she says. "It requires ten to twenty times more oxygen than resting."

While bar-heads are native to Asia, Meir found some to study at a bird reserve in North Carolina's Sylvan Heights Bird Park, where rare and endangered birds from around the world are bred. There she became, she says, a modern-day Mother Goose. Newborn waterfowl imprint on the first creatures they see, so Meir traveled to the park before the birds hatched, and "like a crazy person," she talked softly and sang to them while they were still cocooned in their shells, hoping to create a bond with the unhatched baby geese. Then, when the birds hatched, she would be the first thing they saw.

It worked like a charm, and for a few weeks Meir became surrogate mom to six tiny, waddling, down-covered bar-head goslings. "We would go for walks, sit around outside, they would nap on me, follow me, and cry for me at night when I left," she says. All this was done so the birds would be comfortable and work with her. "A wild animal that doesn't like humans is going to be stressed, and so its heart rate is going to be really high, and it won't exhibit normal physiology," she explains. "If they are comfortable with you, though, you can train them to the equipment." The babies were fitted with tiny fake oxygen masks so they would get used to them during studies.

Research equipment fastened to geese would never survive the perilous flight over the Himalayas. So if the bar-heads couldn't go to the mountain, Meir would bring the mountain—at least a virtual version of it—to them. She packed the gaggle of geese into dog kennels, flew them to Seattle, and from there drove them to her lab, which at the time was at the University of British Columbia. She fitted them with masks that measured their oxygen intake and carbon dioxide output as they flew in a wind tunnel that mimicked the howling gales in high mountain skies. Some of the oxygen the birds would breathe was replaced with nitrogen to simulate the thin air at high altitudes.

So what allows a bar-headed goose to accomplish its incredible feats and not freeze to death or get hypoxic? It's a whole suite of adaptations, Meir says. The wings of bar-headed geese are larger in proportion to their overall size than those of other birds, and the bigger the wing proportionally, the more loft it creates, which reduces the amount of energy the bird has to expend to stay in the air. Feathers keep the body warm by trapping and holding the heat the bird generates. Pectoral muscles power all bird flight, but bar-headed goose pecs have more capillaries around the individual muscle cells than those of other birds, which channel more blood to fuel them.

The entire system that takes oxygen from the tip of the goose's beak to the cells throughout the muscles is more powerful in these geese than in other birds. The respiratory system of birds is larger than that of mammals by a factor of four, which means they take bigger and more effective breaths. Bar-heads also store air after it passes through their lungs and then breathe it again, squeezing more oxygen out of it. And when the oxygen is absorbed through the lungs into the blood of the bar-heads—and penguins as well—the hemoglobin in their blood becomes enriched. This means they can transport much more oxygen, the precious fuel that powers flight muscles. "It's all quite extraordinary," says Meir.

These adaptations occurred over many millions of years, and how they developed in the bar-heads is particularly fascinating and may explain why their physiology is far more advanced than that of any other bird. It's extremely rare for birds to migrate above four thousand feet, and most travel at far lower altitudes. Charles Bishop, who studies the birds at the University of Bangor, proposes that way, *way* back when, bar-heads flew the same migratory route that they fly today, from Tibet and Kyrgyzstan to southern India. But that was pre-Himalayas—that is, before continental plates collided and caused the birth of the massive mountain range some seventy million years ago. As the mountains slowly rose, the birds simply kept migrating along their accustomed route, flying just a tiny bit higher with each passing century. As the Cenozoic era slowly ticked by, the bar-head painstakingly earned its extraordinary high-altitude physiology.

Low-oxygen environments occur at the bottom of the world as well as at the top, and for her investigation into the diving physiology of the emperor penguin, Meir and her adviser, Paul Poganis, a long-time penguin researcher, traveled to McMurdo Station, a research facility in Antarctica. Emperors are the largest of all penguins, growing up to nearly ninety pounds and reaching nearly four feet tall. They have a massive white front, a black back, and a splash of bright yellow around the head and neck. The female lays a single egg, which the males tend, sheltering it between their feet from the fierce Antarctic climate. They migrate in the extreme environment at the bottom of the world as well, though not by flying, but by walking, some fifty or sixty miles in single file in below-zero temperatures and hurricane-force winds, a journey made famous in the 2005 film *The March of the Penguins*. But it is the penguins' deep dives that really set them apart from the rest of the animal kingdom.

In the Antarctic whiteness atop an ice floe near McMurdo, scientists maintain a pen made from wooden corral fencing that they call Penguin Ranch. There, Meir and Poganis scour the waddling flocks of penguins for two nonbreeders they can kidnap for their learned poking and prodding. Once they find a couple of likely candidates, a chase is afoot. "We land in a helicopter or take a snowmobile and get close," Meir says, describing the scientific penguin-snatching about to unfold. As the birds scramble along in their long lines, the researchers sneak up behind the one they want and quickly and quietly herd it away from the others. "We have a shepherd's crook, a long rod with a curved end, and get behind the penguins and pull them back on their feet a bit," she says. "You don't want to knock them down, though, you want to keep them upright, because if they get down on their belly and start tobogganing away you'll never catch them. Once you throw them a little off balance, you quickly throw the crook away, slide in on your knees, and give them a big bear hug around the chest." Taking great care to stay away from the wings—"they batter you, and the wings are really hard, like little two-by-fours, and even though we wear big padded parkas we still get bruises"—Meir and her cohorts stuff the birds into a specially constructed wooden box with lots of padding. In the blackness, the birds quickly calm down.

With the penguin tucked inside, the box is loaded onto a sled and either pulled behind a snowmobile or flown by helicopter to a surgical hut where the birds are anesthetized. As they sleep on a table, miniature equipment is implanted, so when the penguins are released into the water, Meir can measure the number of heartbeats during the dive and oxygen levels in their blood.

Once the penguins come to, trussed up in their scientific instruments, Meir and the others cart them back out onto the ice, where they are released into the pen built at the ranch. The birds are dropped into one of two holes in the thick ice, the only open water for miles around. That means they must resurface at these holes, and Meir can fetch the data devices.

We usually think of birds as being most at home while soaring through the sky, not slicing through the water deep beneath the surface of the ocean, but penguins descend to depths of 1,500 or 1,800 feet and stay there for almost half an hour, all on a single breath. Human divers with oxygen tanks, by contrast, are advised to go no deeper than 130 feet. There is a "whole suite of physiological responses" to enable the penguin's incredible and precise management of oxygen, says Meir, and the birds display masterful control of their bodies. They easily regulate their heart rate, for instance. "If you can decrease your heart rate you can hold your breath longer," Meir says. "It occurs in all animals, but in a penguin in a dramatic fashion." At rest, a penguin has the same heart rate as a human, about 70 beats per minute. But before and after the dive the penguin's heart rate goes way up. Meir sounds amazed, still, as she recounts the data. Human hearts max out at 180 beats per minute. A penguin's heart soars as high as 250 bpm. Their tissue and organs become deeply saturated with oxygen—money in the bank. During the dive, however, when the penguin needs to conserve oxygen, it dials its heart rate way down, to as low as 3 bpm, and sustains a rate of 6 bpm. "Imagine," she says, "the heart is only beating thirty times in a five-minute period!"

Meir hopes to answer more questions about the penguins' superb diving capabilities. No one knows, for example, how the birds go so deep without being crushed by the force of the water or why they don't get the bends, which creates gas bubbles in the blood and often cripples and kills human divers. "They are the best diver in the bird world, diving deeper and longer than any other," she says. "So it makes sense that they have the most dramatic physiological response. We would be unconscious, and they are still down there, swimming around, foraging, and catching their prey."

Penguins are also able to cut off the flow of blood to some organs during the dive in order to conserve oxygen, and when that flow is restored—through a process called reperfusion—they do it without damaging the organ. When humans experience this phenomenon—when a kidney is clamped off for surgery, say, or during a heart attack or stroke—substances called free radicals are generated, and they cause serious tissue damage. Penguins have found a way to block the formation of free radicals, and understanding how they have accomplished this could lead to significant advances in human medicine. "Do penguins have some kind of free radical scavenging system?" wonders Meir. "If so, is it something we can manipulate pharmacologically or through gene therapy that could one day be used medically by humans?"

Once the data on the penguins is collected and the gear removed from their veins, the birds are released back onto the sea ice and waddle off to rejoin their mates.

What does the future hold for the world's most extreme performers?

There are serious problems in penguin land. Eleven of the world's eighteen species are in decline, and Meir's emperors face the most serious threat. The largest emperor colony is forecast to shrink from three thousand to as few as four hundred breeding pairs by the end of the century. That's because Antarctica is the fastest warming part of the globe, and warmer temperatures are reducing the volume of sea ice. Microscopic algae grow in profusion under the floes and become food for the shrimplike krill that the penguins eat, so less ice means less food. A warmer world also means less snow and more rain. Newborn penguin chicks are adapted for temperatures of fifty or sixty below zero, but not rain, and because their feathers don't shed water, hypothermia has increased infant mortality among them. They may be highly adaptable, but their world is changing too fast for them to survive.

Bar-headed geese are not a threatened species, but migrating birds as an overall group are especially imperiled. (Some two-thirds of the species in North America migrate.) More than thirty migratory species are critically endangered, and many more are in trouble. And they are particularly at risk because of their perilous travels. While birds with small ranges can be protected in a park or preserve, migratory birds are exposed to a gauntlet of threats as they move across the globe, from hunting to collisions with structures to loss of habitat. And climate change is making it all even more unpredictable.

A poster chick for the problems migrating birds face from a warmer world is the Hudsonian godwit. This long-legged brown-and-dun-colored shorebird owns the title of the world's longest non-stop migrant in the world, some six thousand miles, from the far south of Chile to the far north of Canada in the northern spring. It, too, has a remarkable physiology. As it readies to leave on its marathon, the godwit shrinks its digestive organs, triples the size of its pectoral muscles, and nearly triples its body fat reserves. Because of changing temperatures in several places along its route, it now reaches its spring habitat in Manitoba later than it used to, and by the time the babies hatch, the insects they rely on for survival are gone. Hudsonian godwit chicks are starving to death in the nest.

These high fliers and deep divers have crucial lessons for us, lessons found in the farthest reaches of the earth. They can help us prevent or treat serious illnesses that result from crippling strokes, heart attacks, and vascular disease. And they are marvels of nature, inspiration for our imaginations and our dreams. But if we want to learn theirs and a flock of other secrets, we must assure their future.

# PART IV

## Birds and the Hope for a Better Future

# CHAPTER 15

Nature's Hired Men:
Putting Birds to Work

The bluebird carries the sky on his back.

—HENRY DAVID THOREAU

Imagine if we could get birds to eat the right bugs in the right place at the right time—to weaponize them, essentially, to protect our food crops. For half a century, from the late 1800s until the 1930s, the federal government actively pursued a policy they called "economic ornithology." Harnessing this bird service proved frustrating. Attracting birds to a certain field to eat a specific pest at the right time of year is akin to herding cats. And measuring their performance proved difficult. Back then, the process of measuring the success rate consisted largely of trapping birds to dissect their stomachs and take an inventory of the contents. Over fourteen years, one researcher estimated, he had gathered "about thirty-two thousand bird stomachs, of which some fourteen thousand [have] been examined." He divided his collection of stomach contents into good, bad, and neutral categories, according to how much of what the birds ate benefited farmers. Much of the time, though, it was tough to sort out what, exactly, was what.

In the early twentieth century, economic ornithologists believed deeply in the idea of squadrons of birds patrolling the farm fields of America. The helpers from the sky were the "farmers' and horticulturalists' hired men," wrote C. D. Howe, Vermont's state ornithologist, in 1915. "They work day and night, seven days a week, thirty days in a month, 365 days in a year. These hired men do not ask for wages, do not ask lodging, and they board themselves. These hired men do not ask for a day off to go fishing or to the ball game; they do not go out on strikes or lockouts, and they do not get drunk on Saturday nights. They are perfectly reliable workmen."

When DDT and other powerful pesticides came along, the federal government abandoned the idea of birds as hired workers; the new methods of pest control seemed to render notions such as economic ornithology quaint and unnecessary.

Now, though, the idea of enlisting one of nature's creatures as a mercenary to battle another, undesirable species is gaining widespread acceptance. Ron Rosenbrand, manager of Spring Mountain Vineyard, a winery on the forested flank of the Mayacamas Mountains in California's Napa Valley, is at the forefront of this movement. They call it "avian agent biological control," and it's a bargain that provides bluebirds with a safe place to live in the vineyards in exchange for their passion for eating grape-killing insects.

Ron is a big guy, friendly and relaxed, especially considering there are a million things out there that could do in his grapes. He is in his mid-fifties, and for many years he managed the Charles Krug winery, one of the largest in this valley, and then was hired by Spring Mountain Vineyard, a European-style winery owned by Jacqui Safra, a Lebanese-Swiss Jewish banker and secretive billionaire whose family controlled the gold trade in the Middle East in the days of the Ottoman Empire.

The Spring Mountain winery is one of the finest in Napa, with a large restored Victorian home called Villa Miravalle, trimmed lawns, and carefully cultivated gardens. Oak barrels of wine are aged in historic hand-dug caves. When Safra bought it, he added three other Napa Valley properties that were each founded in the 1880s; together they comprise 845 acres of forests and vineyards. Pinot noir, Bordeaux, cabernet sauvignon, chardonnay, and other grapes are planted on a quarter of those acres, in 135 blocks interspersed throughout the forests on the edge of the town of St. Helena. The wine they make here is highly regarded, especially the cabernet sauvignon.

In 2003, Rosenbrand walked down one of the rows of head-high vines in the vineyard and noticed critters called vine mealybugs on his grapes. They are small, flat bugs, covered with crumbly white wax, whose excrement creates a sticky slime on the fruit that vintners call honeydew. It makes a mess and is a platform for the growth of grape-killing mold. The bugs can also transmit viruses to the grapes. Vigilance is essential to wine grape growing, and immediately the county agriculture commissioner was all over Spring Mountain's mealybug problem, as well as that of other nearby vineyards that had the same pest. They were told they needed to spray to prevent the bugs' spread. "I was forced to sign a compliance agreement and to spray the grapes with some of the most toxic insecticides I've ever worked with," Rosenbrand told me as we tooled through the winery's forests on dusty dirt roads in his pickup truck, and he pointed out the different vineyard blocks and the 150-year-old olive trees that grow on the property. "It didn't work." The mealybug outbreak worsened because the bug spends much of its life under the bark of the grapevine, where it's sheltered from the toxic rain of pesticides. Because of a previous success using insect predators to control spider mites, Rosenbrand asked the commissioner if he could introduce two parasites that kill mealybugs, called anagyrus and cryptolaemus. He got the go-ahead and ordered some from an insectary, a biological pest control supplier. Lo and behold, the release of the two predators was a tremendous success and the mealybugs went away.

Then in 2005, the blue-green sharpshooter, another vineyard terrorist, showed up on the vines at Spring Mountain. These bugs don't harm the grapes themselves, but part of their life is spent in the wild vegetation along creeks where they pick up a bacterium that causes Pierce's disease, which they carry with them when they fly to the vines. As the sharpshooters feed on sap in the plant stems, they inject the bacterium into the plant. Because sharpshooters live beneath the leaves of the grapevine, chemical sprays need to be applied heavily across the vineyard. Rosenbrand sought to relive his previous triumph, and asked around about insect enemies, but a friend, an organic vintner, said there wasn't a good one for sharpshooter biocontrol. "So the organic guy said, 'Try bluebirds,'" recalls Rosenbrand. "And I said, 'Where do I buy bluebirds?' And he said, 'No, you don't buy them, you put up birdhouses.'"

There are three bluebird species on this continent: the eastern, western, and mountain bluebird. America's bluebird army is healthy now, but they very nearly disappeared in many places, thanks to one eccentric devotee of William Shakespeare.

In 1871 an oddball organization called the American Acclimatization Society formed to import various species of flora and fauna from Europe to the United States, "as may be useful or interesting." The head of the society, a New York drug manufacturer named Eugene Schieffelin, thought it would be a grand idea to introduce all of the birds mentioned in Shakespeare's works. The most successful of the introductions was the European starling. A hundred of the birds were released in Central Park in 1890, and then a hundred more were released a year later. Sixty years later, there were some 200 million starlings in the United States, and today they are found coast to coast.

They are mentioned in Shakespeare's *Henry IV* by a character known as Hotspur (for his impulsivity) who plans to use the birds' obnoxious squawks to drive the king mad as revenge for not coming up with the ransom to free his disloyal brother-in-law Edward Mortimer. "I'll have a starling shall be taught to speak nothing but 'Mortimer' and give it to him, to keep his anger still in motion," Hotspur says.

As the starlings spread out across the country they competed with bluebirds for nest-building cavities, destroying the bluebirds' eggs and preying on fledglings. Combined with the human destruction of bluebird habitat and the use of DDT, the starlings nearly wiped the bluebird off the map.

A 1976 book, *The Bluebird: How You Can Help Its Fight for Survival,* sounded the alarm about the plight of the bluebird, whose numbers had plummeted by 90 percent in some places. A national campaign to construct birdhouses, though, returned the bluebird's population to healthy levels. Providing a safe home does wonders for a population—predators take fewer of them, parasites are lower in number, and more chicks survive to become fully grown.

In 2006 Rosenbrand bought ten bluebird houses from the local Audubon Society at thirty bucks apiece. There's not much to them—just a rough wooden box with a peaked roof and the right size hole, one and three-quarters inches across. "The main thing is that the diameter of the hole be a specific size so the bluebird feels safe," Rosenbrand says. "Mother Nature has taught them that." The perfect size hole admits adult bluebirds but keeps larger predatory birds from wriggling their way in and taking over the house. Snakes and raccoons also climb up the posts to make a meal out of the baby birds, so plastic collars that prevent predators from climbing or slithering up to the house are also important. After Rosenbrand had everything in place, all he had to do was wait for the birds to show up. It didn't take long. "I started seeing bluebirds that first spring," he says. "I thought, 'Wow, this is a good start.' "

Meanwhile, employees put up small, sticky yellow cards around the vineyard to trap bugs and monitor their numbers. Before the houses went up, they were catching twenty-five to thirty bugs per card over a week or ten-day period; after the houses went up, those

numbers tapered off. "The next year we built more birdhouses and
noticed more of a decline in sharpshooters," Rosenbrand says. "I
thought to myself, 'Is this really working?!'" It sure seemed to be,
and by 2007, Spring Mountain had 125 boxes. In 2009, Julie Jedlicka,
a specialist in environmental studies at the University of California,
Berkeley, came to the vineyard to begin a study of bluebird pest control. Rather than dissecting hundreds or thousands of bird stomachs, Jedlicka's molecular scatology study at Spring Mountain
examined the DNA in bird poop to learn what the birds were gobbling. And the news was good: She found sharpshooter DNA.

The number of sharpshooters stuck to the yellow cards continued
to decrease, so the next year they added more houses, bringing the
number from the original few dozen to five hundred, and they had
fewer sharpshooters still. Building continued, and when I visited,
there were nine hundred birdhouses fastened to fence posts across
the vineyards. Birds occupied about 15 percent of them. When I visited I saw a dozen cobalt-blue birds flitting in the sun across the
vineyards. They sat on fence posts, "ground sallying"—flying to
land to grab an insect and then leaping up to fly back to their perch
to eat it. Or they would "hover hunt," remaining stationary in the
air several feet above the ground, then swooping down when the insect prey showed itself. "They have virtually annihilated the sharpshooters," Rosenbrand told me. "They are gone from the traps. And
we don't see Pierce's disease. We're very, very happy."

What swayed a no-nonsense vineyard manager to rely on something like birds for his highly valued crop? Rosenbrand had a come-
to-Jesus moment for biocontrols after he was diagnosed with colon
cancer at age forty-eight, and a colleague and good friend of his was
diagnosed with prostate cancer. Both recovered, but Rosenbrand
wondered if those diseases weren't the result of a lifetime spent using
industrial-strength pesticides and herbicides. "We weren't as careful
as we are today," he said.

While birds will eat pests no matter what people do, a study by
Jedlicka found that providing the boxes concentrates the effect. She

compared vineyards with nest boxes in the Mendocino Valley to those without, and found that adding the boxes resulted in 50 percent more species of insect-eating birds—including tree swallows and violet-green swallows, which also take possession of some of the houses—and the density of all insect-eating birds quadrupled. There is concern in Napa about the possible imminent appearance of another deadly pest with a similar name, the glassy-winged sharpshooter. I saw signs with pictures of the bug that looked like WANTED posters, asking people to report them if they saw them. But Rosenbrand has a crack team of bluebird predators now and he expects they will eat those as well.

Jedlicka is a cautious scientist, and all cautious scientists want more research. But she thinks the approach has serious promise, though there are lots of angles to work out to make sure target species are taken, and to enhance the effect. And there is additional reason to be cautious about harnessing bird services—the law of unintended consequences.

Crimean-Congo hemorrhagic fever is a virus carried by ticks that showed up in Turkey in 2002 and spread widely, causing fever, headaches, and severe muscle and joint pain and, alarmingly, killing more than three hundred people. When a 1992 study found that an African bird, the guinea hen, ate ticks in the wild, scientists wondered if they might help prevent the spread of Lyme disease. Someone in the Turkish government read about this theory and hit upon the idea of a feathered strike force. In 2010 they introduced hundreds of thousands of imported guinea fowl around the country to take on the ticks. Entrepreneurs started selling them to nervous landowners. The urge to buy the birds and deploy them was so strong and the belief in their powers so great, it felt cultlike, Sekercioglu, the Turkish bird biologist I visited in Utah, told me. He warned officials against it. "I told them they should stop immediately because they might be doing the opposite of what they intended," he said. Followup research, in fact, called the earlier study into question and showed that the birds did not eat many, if any, ticks. Instead, ironically, a single bird carries hundreds of ticks in its feathers, and the guinea fowl spread the disease to new locations.

Next on the list of pests Rosenbrand is using "nature's hired men" to deal with are voles, tiny rodents that burrow underground. "They've gone crazy the last few years," he told me. Voles gnaw at the roots of the grapevines, which kills them. "Controlling them organically is difficult, especially because we have a cover crop of grass between the rows of grapevines, which allows the voles to hide," he said. He has employees cut the grass short "so the hawks can get at them, and owls, and we bought some snakes and released them." The raptors also take linnets and starlings that feed on the grapes.

While it is still being worked out here, rodent control has been harnessed and enhanced in a number of other places. In a swords-into-plowshares scenario, Amir Ezer, a former Israeli naval officer, photographer, and bird-watcher, distributed thousands of empty ammunition cases to be made into nest boxes for owls on Israeli farms. A single pair of barn owls—white-and-tan birds with curious heart-shaped faces and dark, slanted eyes—will swoop in and scoop up as many as five thousand rodents a year. Some farms have found them so effective they have eliminated poisons. "From the moment the barn owls nest, the damage stopped," says project coordinator Shauli Aviel. Because the barn owls hunt by night, and crops were vulnerable during the day, the project added boxes for the day-hunting kestrels. It was a smashing success, and more than 25,000 acres of farmland have been pesticide-free for a decade.

Still, the subject is poorly researched and the impact of owls might be even greater than we know. Mice tracked as barn owls fly overhead, for example, show nervousness—a phenomenon known as "apprehensive foraging"—and reduce their time in the fields by a quarter. This adds to the positive impact from predators that usually goes unnoticed.

Although the Spring Mountain winery is not certified, it is de facto organic, thanks to the bluebirds and their ilk. "We've not sprayed insecticides here in four years, and we stopped using herbicides," Rosenbrand told me happily. It's meant more work in a business that already relies on a lot of physical labor—each grapevine is touched by workers fifteen to seventeen times a year, and organic means even more touches. "But four years ago you couldn't find an earthworm, and now earthworms are everywhere. Everything is alive again, and before that it was sterile and dead."

There is another way wine country harnesses bird services, one that is particularly elegant. Vahé Alaverdian, the owner of a company called Falcon Force, based in La Crescenta, California, is a falconer who keeps a couple of dozen striking birds of prey as hired labor: peregrine falcons, Barbary falcons, Harris's hawks, gyr/peregrine falcon hybrids, and an unusual and stunning bird, the Aplomado falcon from Peru. I clicked through the pages of the Falcon Force website and found photos of the birds listed under the heading Our Staff. When I called Alaverdian to ask about his team, he told me that birds are "just like puppy dogs; they each have their own personality. And peregrines are probably the sweetest falcon to work with."

Alaverdian travels the West Coast with his predators, practicing falconry-based bird abatement. He doesn't just address crop damage. His year starts in San Diego in the summer, when he heads to Sea World and sets up shop near Manta, a stomach-churning roller-coaster ride shaped like a manta ray that overlooks Mission Bay. Gulls flock around the ride and roost on the rails, and people have been whacked in the face by the birds as the coaster comes shooting out of the gate at eighty miles per hour. Alaverdian flies his falcons when the rides are operating, releasing them to circle high above the towering rails. The gulls disappear when Falcon Force is on patrol, and every once in a while a passenger on the Manta gets the added thrill of seeing a falcon take out an unlucky gull.

Alaverdian and his team also work the wine grape, cherry, and blueberry harvests in California, Oregon, and Washington. If it's a vineyard, the key time is just as the grapes begin their first blush, turning from green to purple and getting sweeter, a process known as *veraison*. "A flock of a thousand starlings, which is not a big flock, can drop into a seven-hundred-acre vineyard, and in a matter of a week it could be a complete loss," he told me.

Alaverdian finds living quarters near the vineyard where he and his birds will be working, and each day for several weeks, in the early morning darkness, the falconer arrives with his birds. "When the light breaks, you are on top of the highest vantage point, and you are scanning for birds coming in," he said. "As soon as that one bird lands, you put the falcon up." That's because the first starling is a scout, "and one bird will bring in the next five birds, then fifteen, then a hundred and fifty, and pretty soon you've got two thousand birds," and all of them are jonesing for newly ripened grapes.

When Alaverdian releases the falcon, it skies up, corkscrewing to an altitude of fifteen hundred feet or so, where it flies for about forty-five minutes. Having scared away the starlings simply by its presence, it returns to the feather lure that Alaverdian swings around his head, and when it lands it is fed a bit of meat. Then another bird is sent aloft, and the cycle is repeated later in the morning when it's still cool, and again later in the afternoon as the heat of the day diminishes. If the temperature is still warm, Alaverdian showers the birds with mist from a spray bottle.

Falconry is a precision occupation. "Every bird has its set weight at which they perform best, just like an Olympic athlete," Alaverdian told me. Too well fed and they will sit lazily on a fence post and become eagle bait. Too little to eat and they might not have the energy to fly. The birds are weighed three or four times a day to keep them at their fighting weight.

For three months Alaverdian spends ten or twelve hours a day flying his falcons and keeping watch over the grapes. Bird abatement with falcons is cheaper and easier than spreading fine mesh mist nets over the vines, and more effective than pyrotechnics or propane cannons. And it's certainly easier than the solution one vintner found:

riding his bike up and down between the rows of grapes, constantly ringing its bell.

After I had spent my day at Spring Mountain Vineyard learning how birds are enlisted in the war on pests, Ron Rosenbrand dropped me off in front of a small historic house on the property to sample some of the vineyard's wine. At the tasting, I found the vineyard's prize cabernet exquisite, both exuberant and seductive, with black cherry and bittersweet chocolate notes, all the more delicious because I knew the grapes had been watched over by tireless bluebirds, those perfectly reliable workers who "do not go out on strikes or lockouts, do not get drunk on Saturday nights."

The take-home lesson of the bluebirds is that we need to understand how natural systems work and find ways to partner with them, not fight against them, which we have been doing for far too long. While it may have seemed that spraying crops with chemicals was easier and more effective, the only way that really works is if people close their eyes to the true costs. While pesticides may have killed a lot of bugs, they also took out a lot of birds, worms, and more than a few farmers and farmhands along the way. Birds show us the way to a more human-friendly agriculture, a place where they—and we—can flourish.

## CHAPTER 16

~~~

# The City Bird:
# From Sidewalk to Sky

Everything has beauty, but not everyone can see it.

—CONFUCIUS

Tales of a deep bond between birds and humans come up again and again in bird lore, and one of the most intriguing and inspiring is the story of Nikola Tesla and his pigeons. Tesla was the brilliant and wildly eccentric Serbian inventor whose work led to the adoption of alternating current, our nation's method of electrical generation, instead of Thomas Edison's direct current. After retirement, Tesla spiraled further into eccentricity, and he became obsessed with the pigeons in his New York neighborhood. In the 1920s and '30s he carried a sack of bird food and cast handfuls on the ground as he walked through Bryant Park behind the New York Public Library. If he whistled, legend has it, the pigeons would flock to him, and even land on his arms and shoulders. On days when Tesla couldn't feed them, he hired a child to take his place.

After moving to the swanky Hotel New Yorker in 1934, Tesla was warned more than once that he could no longer feed and attract the birds to his window, because it made such a mess, but he couldn't quit them and kept at it. "Sometimes I feel that by not marrying I made too great a sacrifice, so I have decided to lavish all the attention of a man no longer young on the feathery tribe," he told a reporter from the *New York World*. "I am satisfied if anything I do will live for posterity. But to care for those homeless, hungry or sick birds is the delight of my life. It is my only means of playing."

Tesla once came across a wounded female, a white bird that he was absolutely mad about. "I have been feeding pigeons, thousands of them for years," he wrote. "But there was one bird, a beautiful bird, pure white with light gray tips on its wings; that one was different. I had only to wish and call her and she would come flying to me. I loved that pigeon as a man loves a woman, and she loved me. As long as I had her, there was meaning to my life." He nursed her back to health. "Using all of my mechanical knowledge I invented a device by which I supported its body in comfort to let the bones heal," he said, and spent more than $2,000 to cure her.

One day the white dove appeared at his window and the Serb knew the end was near for his pigeon love. "As I looked at her I knew she wanted to tell me she was dying," he said. "And then, as I got her message, there came a light from her eyes—powerful beams of light."

"This devotion to his pigeon feeding task seemed to everyone who knew him like nothing more than the hobby of an eccentric scientist," wrote John J. O'Neill in his book about Tesla, *Prodigal Genius*. "But if they would have looked into Tesla's heart, or read his mind, they would have discovered that they were witnessing the world's most fantastic, yet tender and pathetic, love affair."

———

The plump, head-bobbing pigeon has been deeply intertwined with human affairs for longer than any other bird. I would even argue that the pigeon may be the closest buddy—even soul mate—from the bird world that humans have. "Pigeons are often a city child's first contact with nature and an elderly person's only friend," is how New York city councilman Tony Avella put it. That might be more important than it first seems. Some scientists and conservationists think that because of the relationship, the charismatic pigeon may be the last tenuous thread between urban humans and nature, and therefore an ambassador to enlist support for protecting the world's biodiversity outside cities.

While the pigeon fan club has many members, there's a whole flock of people who hate them with equal fervor and would love nothing more than to rid the world of them. I wanted to know more about these two sides of the pigeon coin, and as I looked for a battle-ground to visit, I was surprised that instead of having to travel to the pigeon-crowded streets of New York or to St. Peter's Square in Rome to find experts and partisans, I found an avian drama unfolding less than an hour from my house, in Butte, Montana.

Unlike any other bird, pigeons thrive in the brick and asphalt heart of cities, and it turns out that Butte is one of the most urban cities in Montana, at least from an architectural perspective. A mining town, it sits atop the so-called Richest Hill on Earth, which was shot through with veins of gold, silver, and especially copper. The glory days of the copper kings who made billions and built a city that rivaled New York and San Francisco in the frontier wilderness of Montana are long over now. The ore played out, and the population plummeted from its peak of more than 100,000 early in the twentieth century to its current 35,000 or so. Many of the historic buildings still stand, but a good many of them are empty, crumbling shells with broken windows and boarded-up doors. And it was here I found Stella Capoccia, an assistant professor at Montana Tech, whose field is something called animal geography and who is refereeing the two sides of Butte's pigeon war.

On a sunny spring afternoon, I met Capoccia for lunch at a local eatery called the Hummingbird Cafe. She's petite, in her thirties, and was full of energy even though she had given birth just a few weeks before. Animal geography is a field new to me. It deals with "the complex entangling of human-animal relations with space, place, location, environment and landscape," according to a collection of research papers called *Animal Spaces, Beastly Places*. Capoccia worked for a time in Africa, studying how animal advocacy groups influence Kenya's wildlife management. Then she came to the American West, landing with her husband in Butte in 2010. It turns out that Butte was in need of the services of a good animal geographer, and she was contracted by Silver Bow County, of which Butte is the county seat, to study possible solutions to the conflict.

Pigeons are at the heart of what animal geography is all about. They are found on every continent except Antarctica, and exist in places most other birds don't, filling urban centers everywhere with their soft coos and slapping wings as they take flight. What would the world's great cities be without these ubiquitous flocks? Cleaner for sure, but nearly devoid of nonhuman life. A flock of pigeons bursting into flight in unison lifts our gaze off the crowded and noisy streets, out of the shadows of buildings, and into the sunny sky above the fray. It's a reminder of the natural world that exists beyond busy, hyperfocused urban lives.

One reason the pigeons are so abundant in cities is that pigeons mate more than once a year, and since they first became domesticated they have had many thousands of generations, which mean that traits prized by humans—speed, size, homing ability, docility, colors, and reproductive vigor—have all been well expressed in pigeons. Two of those traits—tameness and reproductive vigor—explain a lot about why today's pigeons flourish in cities. The testes of the males, because they were selected for mating prowess, are substantially larger than those of their wild brethren.

In his book *Feral Pigeons,* the ornithologist Richard Johnston argues that the city pigeon and the wild rock dove are similar but not identical. We have taken a wild cliff dweller at home among people and selected its genetics over thousands of generations to create an urban pigeon that can survive in the most inhospitable of places, "a creature of both wild and domestic ancestors that has in a sense become the best of all possible pigeons, with capabilities transcending those of either ancestor . . . one of the masterpieces of nature."

Officially the pigeon is a rock dove, in Latin *Columba livia* (literally, "blue [or blue-gray] dove"). Wild pigeons are usually a nickel gray, with black bars on their wings and a spray of iridescent feathers the color of an oil sheen around their neck. They belong to the same family as mourning doves, turtle doves, and passenger pigeons. The ecological niche created by the roads, alleys, cobbles, buildings, and bridges of the world's cities and suburbs are not all that different from the treeless rock cliffs of southern Asia where the species originated millions of years ago. (Cliffs are where they developed their "burst" flying style, that sudden upward takeoff.) They didn't eat old French fries and stale bread crusts in ancient Eurasia, of course, but many city pigeons still eat a native diet of grain and seed, in addition to human leftovers.

Among the first birds to be domesticated, between ten thousand and five thousand years ago, were the wild rock doves, who came in from the cold to live with people along the Mediterranean. They probably first cozied up to farmers, who grew the grain the birds loved. Cities then were built of mud and rock, and doves may have been kept and raised as food, or they simply moved on their own into the burgeoning urban areas, not seeing it as all that different from their wild cliff dwellings. It worked for both parties—the pigeon had a place to live free from most predators, and people had an animal companion they could harvest.

Rock doves have been more than a friend and a quick fricassee, though. They are a sacred symbol to many religions, from the dove of peace to the Holy Spirit, which appeared in bird form before Jesus as he wept and suffered on the cross. "And the Holy Ghost descended in a bodily shape," writes Luke in the Gospels, "like a dove upon him, and a voice came from heaven, which said 'Thou art my beloved Son; in thee I am well pleased.'" The dove was a symbol for Hercules and a central character in flood myths, such as the sentinel bird that the biblical boat builder Noah releases as the waters of the Great Flood begin to recede. As the Hebrew Bible tells it, the dove returns to the boat because there is no place to land, but after being released a second time, the dove appears with a green olive leaf in her beak, a sign from land that the waters are finally receding. Seven days later, when the dove is sent out yet again, it never returns, indicating the flood has passed.

Throughout history, people around the world have kept and bred pigeons for food, sport, and their beauty. Charles Darwin, himself a pigeon fancier, wrote of the sixteenth-century Mughal emperor Akbar Khan, "never less than 20,000 pigeons were taken with the court. . . . His majesty, by crossing the breeds which method was never practiced before, has improved them astonishingly."

The birds were raised throughout Europe and Asia for centuries, and many large homes and farms had dovecotes, aka pigeonaires, aka living pantries, where the birds roosted at night and flew out during the day to forage. Some wealthy pigeon-lovers built large, fanciful structures to match their homes that housed thousands of birds. To pay the rent the birds were harvested, and their rich droppings dug into gardens and farmland.

Pigeons began their conquest of North America in Halifax, Nova Scotia, in 1606, when they were brought to North America from Europe by the explorer Samuel de Champlain, who used the birds as a source of food during his explorations.

A pair of pigeons breed about a dozen babies per year. Young pigeons, called squab, are harvested between four and five weeks of age. Squab are near adult size, but because their flesh is youthful, their meat is far more tender, with a silky, rich texture that falls off the bone, something like dark chicken meat. Squab is still highly regarded and eaten around the world, especially in France and China.

Charles Darwin might not have arrived at his theory of evolution were it not for pigeons. He began breeding "fancy pigeons" and visiting London pigeon clubs. As he worked with them, he realized that these very extraordinary-looking birds, whether the English pouter with its oversize breast or the oriental frill-back with its poodlelike curlicue feathers, or the short-faced tumbler, which somersaults as it flies through the air, were all descended from common, ordinary rock doves. Human breeders selected for, and exaggerated, a range of different, desirable traits over centuries to create new, very distinctive, sometimes outlandish pigeon breeds. In fact, human selection so dramatically changed the look of pigeons, Darwin wrote, that they looked not only like a different species but even like a different genus. He had a revelation: If genetics are so plastic that humans could select for such traits and come up with such dramatically different outcomes, nature might do the same, selecting for traits based on factors such as weather, disease, and predators.

The National Pigeon Association, formed in 1920, today recognizes more than 450 distinct breeds, with hundreds of variations on each, all engineered by humans from the plain gray rock dove. And new breeds are still being created.

Many people, I wager, are so consumed by their everyday regimens in cities that they don't even perceive these birds. But there's an important story behind this tough little streetwalker scrambling for crumbs in the gutter. Pigeon intellect, for example, has been greatly underestimated. These birds learn much the way human children do, and their vision and perception are uncanny. They can tell the difference in painting style between Monet and Picasso. Their vision is so good that the Coast Guard put them to work years ago on a project called Operation Sea Hunt. Trained to spot orange objects by receiving a pellet or two of food when they did, the birds were mounted in Plexiglas domes under rescue helicopters. In choppy seas the pigeon sentinels performed much better at spotting shipwreck survivors in life vests than did human searchers.

Pigeons are the original working birds, and people have long partnered with them because of their unerring and abiding love for home. A champion homing pigeon can be released six hundred miles from home and, zipping along at more than sixty miles per hour, return within a day. This ability has made pigeons singularly useful to human communications. As far back as the fifth century B.C., a complex long-distance communication system was established between cities in Persia and Syria with pigeon couriers. In ancient Rome, results of the Olympic Games were spread via pigeon. Before the telegraph, the Rothschild banking family set up a network of pigeon lofts to house the birds in order to rapidly disseminate financial news. Paul Julius Reuter founded the news agency that still bears his name in the mid-nineteenth century first using carrier pigeons to close the seventy-eight-mile gap between Aachen, Germany, and Brussels, Belgium, with breaking news. During both World Wars, soldiers on both sides carried baskets of messenger doves into the trenches. Carrier pigeons were so essential they were deemed official members of the military and often honored for their bravery. And when German spies sent their pigeons back to Germany, the Brits used peregrine falcons to hunt them down.

The most famous war pigeon is Cher Ami, whose name means "dear friend" in French. In October 1918, during World War I, a U.S. battalion made up of more than five hundred men was surrounded by German troops at Grandpré, France, during the Battle of Argonne. Their numbers were decimated, first by the Germans, then by "friendly" artillery fire. Unable to get word to their own forces about their predicament, they started releasing their flock of carrier pigeons one at a time. And, one at a time, the birds were shot down by German fire. Down to their very last bird, Cher Ami, the desperate battalion dispatched him into the air with a note rolled up inside a small canister fixed to his leg that read: *We are along the road parallel to 276.4. Our own artillery is dropping a barrage directly on us. For heaven's sake, stop it!*

As the Germans spied Cher Ami flapping furiously through the smoke-filled air, intent on his intelligence mission, they opened up on him. Though one leg was hit and a bullet lodged in his breast, the fearless bird defied the deadly fusillade. Just sixty-five minutes later he glided into his loft at division headquarters, twenty-five miles away. The bombardment was immediately halted. Cher Ami's heroic journey saved the last survivors of the Lost Battalion, 194 soldiers, and for that he was awarded the Croix de Guerre. Cher Ami died in June 1919 from his wounds and now sits stuffed atop his lone leg at the Smithsonian Institution's National Museum of American History in Washington, D.C.

The era of the message-carrying homing pigeon is not over. The People's Liberation Army of China, one of the largest military forces in the world, recently created a bird corps—ten thousand pigeons to carry messages in case its high-tech communication systems break down.

Yet for everyone who sees in the pigeon a friend and ally, there is someone who sees what Woody Allen disparagingly called a "rat with wings," and not without reason. Though scientists say the risk of pigeons passing disease to humans is infinitesimally low, one bird produces twenty-five pounds of waste a year, and though it is a great boon to farmers, the mixture is so acidic it eats through paint, and damage to buildings in the United States from pigeon poop is pegged at over a billion dollars a year. City dwellers have tried various tactics to get rid of them, from introducing pigeon-eating falcons to smearing a sticky gel on windowsills to prevent roosting, to shooting catapults of thick-roped net over the flocks so that the birds may be hauled away and asphyxiated. Then there's the old switcheroo—workers sneak in when pigeons leave the nest and replace their real eggs with artificial ones that never hatch.

The very best way to reduce pigeon numbers, Capoccia says over lunch, "is the elimination of their nest and feeding options." So in Butte, the first step toward decreasing the pigeon population is asking building owners to board up holes in the myriad empty buildings. It sounds like an easy fix, but the large number of absentee landlords makes the task difficult. Keeping people from feeding birds is second on the list, though putting this into practice is also more difficult than it might seem. There is no shortage of people who feel morally obligated to feed pigeons. On the edge of uptown Butte, Capoccia points out a small brick miner's shack painted pink, with pigeons sitting on power lines and rooftops nearby. "An elderly woman lived there and used to feed the pigeons every day," she says. When the woman died she bequeathed the task to her son, which he dutifully carries out. Hundreds of birds still gather here to await the nightly tossing of the bread, although it's against municipal law.

Some cities, though, are experiencing success in reducing their pigeon populations. London's Trafalgar Square was once, famously, a pigeon heaven, with vast seas of birds swarming around the ankles of tourists. In spite of protests from people who enjoyed the birds, the mayor banned vendors from selling corn to feed them, and Harris's hawks were brought in to hunt them. As a result, pigeon numbers dropped from several thousand to virtually none.

The effort in Butte isn't to wipe out pigeons completely—just to reduce their numbers and the damage they cause. The wishes of pigeon lovers to have the birds around should be honored as well, Capoccia says, so it's good to increase tolerance for the birds in those places where they can live without being a nuisance—on the old towering metal head frames that once lowered the miners into the shafts beneath the city, for example, which still dot Butte's urban landscape. It's important that people in cities maintain a connection to nature, she believes.

A definitive text for understanding the deep emotional nature of the pigeon-human relationship is the 2013 book *The Global Pigeon* by Colin Jerolmack, an assistant professor of sociology at New York University who traveled the world to study the deep bond between urban rock doves and city dwellers. Some of his studies looked at people who fly the bird for sport. By setting a bird free on a four-hundred-mile marathon and then having it return, pigeon fliers, Jerolmack writes, are "vicariously engaged in a dramatic struggle against the powerful and hostile forces of nature, without leaving their roof." When, and if, the bird returns home, it's a "magical, miraculous triumph of nurture over nature."

Pigeon flying also provides bird fliers with what he calls a "social self," a sense of identity in a community, and often a link both to the past and to other cultures. In Berlin, Turkish pigeon fliers carved out their "Turkish space" with their hobby and reinforced ethnic bonds. An elderly Italian American pigeon racer Jerolmack profiled, Carmine Gangone, forged a connection to young Hispanic and black men who, new to his neighborhood, fell hard for the birds and wanted to learn the sport. Pigeon racing has proved itself as a way, then, he writes, "to transcend race and ethnicity."

But what about the birds, I wondered. Is this love affair requited? Do they enjoy not just our food but our companionship? No one can provide a definitive answer, at least not with any scientific authority. But not long after my trip to Butte, a friend in Helena, twelve-year-old Tara Atkins, told me a tale about her pet pigeon, named Foresta. Born and raised in the Elkhorn Mountains, well outside of Helena, Foresta had never been into town. The bird escaped one day, and Tara and her parents couldn't find her. They left for town with the bird on the loose, and they worried about her. Later that day Foresta showed up at Tara's school during recess, alighting on a teacher's bald head before she was captured. Somehow Foresta knew where to find Tara, though she had never been anywhere near the school. The story was written up by the local paper and went global on newswire services.

Is pigeon love—feeling a connection to these birds—just an eccentricity of some people? Or does it exemplify a bond we have with birds? With all animals? Is it an inherent longing, a deep hunger for nature? Are our feelings for birds something fundamental in the human condition that, living in the noise and haste and stress of cities, we've mostly lost the ability to access or perceive? There's a lesson, I believe, in the story of a couple of pigeon lovers in New York I spoke with, Eugene and Kaoria Oda, whom I heard about from a group called People for the Preservation of Pigeons. "We weren't really interested in pigeons, or any kind of animal at all," Eugene said as he told me his story. "Then a pigeon couple started making a nest on the AC unit and it caught our attention. All of a sudden we started looking for pigeons. We always walked past them, we always knew they were there, but we never actually saw them. We started looking at them. We started thinking, 'Oh, what charming creatures they are.'"

Their bird switch really flipped when the Odas noticed a pigeon on the street in Astoria, Queens, that appeared listless and perhaps ill. "At first we were scared to pick it up; we had never touched a pigeon before," Eugene told me. "But my wife got really brave." They brought it home, and soon they took in another injured pigeon. Eugene started volunteering at the Wild Bird Fund to learn how to patch up their growing collection of injured pigeons. And he started his own rehab operation in their apartment, turning their only bedroom into an ICU while he and Kaoria slept in the living room. They keep about twenty birds at home, treating them with medicine, tiny slings and splints for broken wings and legs, and sometimes just bed rest. They name the birds they bring home alphabetically, and in 2012 they went through the alphabet eight times. If a bird recovers, they free it. Adults get returned to the place where they were found, though with babies it's different—they get a soft release. "I spread some birdseed on the ground, so a flock gathers there, and you let the baby watch for a while," Eugene said. "You do that every day and after five days they start trying to get out. They are kind of scared at first, but eventually they join the flock."

This love for the urban pigeon, some scientists think, may be far more than just a curiosity. It may play a key role in stanching the disappearance of global biodiversity. The Pigeon Paradox is a theory that proposes that the pigeon, the most charismatic and accessible of urban species, is vital to protecting all of the world's biodiversity. Because half the world's population (and 80 percent of Americans) live in cities, it's essential to keep an experience of nature alive, research shows, because it encourages environmental sensitivity. "The future of conservation depends on urban people's ability to experience nature," the authors of a research paper titled "The Pigeon Paradox" write. "The complex coos of doves can be soothing and the lives of doves can open the door into a broader interest in wild nature."

The tale of Eugene and Kaoria Oda is a parable for our time regarding our relationship to the natural world and our longing to reconnect with it. One day the couple serendipitously awoke to the glory of pigeons and other birds, and this new transcendent relationship changed their lives. But just being in close proximity to a flock of pigeons isn't enough. That's a starting point, but the real battle to establish a connection to nature is in large measure within ourselves. We must change the way we perceive the natural world. We need to get out into nature, learn more about it, and find new ways of seeing it, the way the Odas taught themselves to see birds. We need a change in perception, to move birds and nature from the background of our lives to the foreground. This is true today more than ever, and birds have something to teach us about that, too.

# CHAPTER 17

The Transformational
Power of Birds

I hope you love birds too. It is economical. It saves going
to heaven.

—EMILY DICKINSON

Fourteen snowy owls have staked out roosts atop fourteen houses in a subdivision in northwestern Montana. These birds are the ultimate in avian gravitas—white, two feet tall, imperturbable, elegant, and decidedly mystical; some call them "ghost owls." They are the same species of owl as Hedwig, who travels with Harry Potter and dies defending him. One of them, a male, as white as a mountain snowdrift, follows our vehicle with his piercing yellow-eyed gaze, his head slowly swiveling as we drive through the subdivision.

Suddenly Denver Holt, the director of Montana's Owl Research Institute, whistles, a long-drawn-out screamlike sound that slices sharply through the winter air. "It's a food-begging call," Holt says of his mimicry. "The young do it when they are hungry." It gets the attention of one bird, who swivels toward us, its Elvis-style heavy lids drooping over its yellow eyes, mildly interested for a second, then returning to indifference. The birds have been sitting on these rooftops for weeks, watching for voles scurrying through a meadow to the south and otherwise doing as they wish with impunity, like a gang that's taken over a neighborhood. The birds have been a big draw, and thousands of people have come to see the spectacle.

The winter of 2011–12 was a banner year in the United States for snowy owls, one of the most widespread mass gatherings ornithologists have ever seen. Arctic birds, snowies don't usually travel anywhere near this far south, and when they do in large numbers it's called an irruption. There are usually irruptions every four years or so, but this one was especially large, and incredibly widespread, extending from coast to coast. It happened again in 2013 and 2014. And no one knows why.

People across the country were smitten. Avid bird-watchers, scientists, and people who knew nothing about birds swarmed to places where the white birds gathered. Many called it a magical experience or a sign—of what, they weren't sure. In the Midwest, the farthest south the birds were seen, people greeted them as celebrities. Ninety birds showed up in Kansas and forty in Missouri, an event absolutely unheard of previously. When five of the birds took up residency at Smithville Lake, near Kansas City, Missouri, it created an "owl jam" as thousands drove there to see them. Newspapers and TV stations gave the birds top billing.

One of the things birds do best is inspire. The soaring eagle, the hummingbird floating in front of a brilliant red trumpet flower or at the feeder, the bowerbird weaving its sculptural nest with coins or bits of foil or other human objects, the kingfisher cleanly cleaving the surface of a river, or the long-legged heron standing immobile in a pond as it waits to spear a fish—all touch us in a deep place and can stop us in our tracks. Owls, though, occupy a special niche. They are the most charismatic of the charismatic—there are even owl cafés in Japan and England where you can share your table with a real live bird. "Maybe it's because they look like us," Holt says, as we watch the rooftop owls. "They have a symmetrical face, with eyes in front and ears on the side of their head, and a billed nose, all of which we have. On top they have fluffy feathers that resemble hair, and a flat face. It's like good-looking people. Some people have these looks that are more attractive to the masses and we stop to look at them with admiration."

These George Clooneys and Scarlett Johanssons of the avian set who perch on the Montana chimney tops have another thing going for them. "White wolves, polar bears, white whales, white buffalo," Holt says. "There is something about white plumage that signifies innocence or purity, the same way we view a wedding dress. People don't flock to see any other animal the way they do white ones. It moves us."

Owls are also associated with sagelike intelligence; in Greek mythology, a baby owl sits on the shoulders of Athena, the virgin Greek goddess of wisdom, and her counterpart in Roman mythology, Minerva. This reputation for wisdom comes partly from their humanlike look and partly from the fact that they are active in the mysterious night.

The charisma of birds is why bird-watching has skyrocketed in popularity in recent years, growing in every demographic, including among children. Nearly fifty million Americans call themselves bird-watchers, and the economic impact this hobby has is huge: Birders spent $15 billion in 2011 traveling to see birds, and $26 billion on equipment.

Bird-watchers spend this kind of money in pursuit of something big, something more than birds. It's not just a pastime, but, for many, a powerful obsession. Take the Big Year, the fiercely contested competition among elite birders to see as many species of birds in a geographically defined area—a state, say, or in North America. The competition began after the first modern field guide to birds was published by Roger Tory Peterson in 1934 and grows more intense each year. The first continent-wide Big Year record was set in 1939 by a traveling businessman named Guy Emerson, who tallied 497 species as he made his rounds (out of 914 species in North America, north of the Mexican border).

These days, the Big Year has morphed into a carefully strategized competition among elite birders who rely on an array of computer technologies and modern transportation. Three birders set a new record of 748 sightings in 1998 and then were bested in 2013 when Neil Hayward, a biotech consultant in Cambridge, Massachusetts, glassed 749 species, ending on December 28 with a great skua, a hawk-size brown seabird that he spotted while sailing along the North Carolina coast. In his blog, *The Accidental Big Year,* he estimated that he spent 195 nights away from home, drove 51,758 miles, was at sea for 147 hours over fifteen days, and flew 193,758 miles on 177 flights through 56 airports. "We could be sitting in a coffee shop and I'd get an email about a sighting and I'm off to the airport," Hayward told the *Boston Globe.* (Just as this book was going to press, at the end of 2016, birder John Weigel had racked up 783 species and set a new record.)

The United Kingdom, though, seems to have more obsessed birdwatchers per capita than anywhere else in the world. It's home to the "twitchers," a particularly fanatical group who make long journeys to see rare birds and presumably go into paroxysms of delight when they see one. They have their own lingo—a *megatick* is a very rare bird, a *crippler* an especially showy rare bird (perhaps because it causes people to stop in their tracks), and *plastic* refers to a bird that has escaped from captivity. The bird world's equivalent of a home run leader goes to John Hornbuckle, who has seen 9,414 types of birds in his life. And he is still adding species to his list.

—

When I put the question of why birds compel us to watch them to Janis Dickinson, an ornithologist at the Cornell Lab of Ornithology who has studied the issue, I got a whopper of an existential answer that I didn't expect. It had to do with fear. Dickinson's lab researches various bird-related topics, from cooperative breeding among the feathered set to the role of citizen scientists in ornithology. She has also looked into the role birds play in the human dimensions of climate change and in 2009 published a paper on the subject. She believes that the denial of climate change in the face of such overwhelming scientific evidence is driven by a deeply rooted fear of our own personal annihilation. Faced with the possibility of a massive global catastrophe, people's own fear of death begins to surface and causes extreme discomfort. By denying the reality of catastrophic climate change, they keep these personal fears at bay. I was taken aback when I read Dickinson's argument, but as I did more research I realized that the ideas in it lie at the very heart of who we are as a species, and moreover, might provide answers not only to why we love birds but to other fundamental questions about the natural world and why we seem hell-bent on destroying it.

In 1974, a cultural anthropologist and sociologist named Ernest Becker won the Pulitzer Prize for his book *The Denial of Death*. Much of Becker's electrifying idea is based on the work of Sigmund Freud, Søren Kierkegaard, and Otto Rank, and it argues that from our earliest time as young human beings we have realized that death is inevitable, which produces a deep, crippling fear of annihilation. Because this feeling is so costly—we can't function well feeling this way—we reflexively repress it, burying it deep inside ourselves beneath conscious awareness. This smoldering, unacknowledged volcano of anxiety, which William James called "the worm at the core," is considered by Becker and these other thinkers to be a condition that all people have at some level.

Humans carry this emotional pain and fear in both body and mind—tensing and numbing muscles and organs, closing off parts of ourselves, distracting ourselves, or self-medicating, lest our fear rise up and be fully expressed. We can live our whole lives managing this terror, and we may do it so well we never realize it's there. Yet hundreds of published studies show it exists, causing or contributing to a raft of psychological and physical problems, including anger, migraines, chronic pain, ADD, depression, decreased immune function, emotional numbness, paranoia, and much more. It distorts our perception of the world, both natural and human. Gandhi recognized it when he said that existential terror is what's at the bottom of our prejudices. "The enemy is fear. We think it is hate, but it is fear."

"It is," wrote Becker, "a mainspring of human activity."

Fear of death, Becker believed, is at the root of the world's most serious crises, from war to torture to genocide, rampant materialism, the abuse of nature, and, Dickinson adds, "efforts to control rather than sanctify the natural." Becker hoped to change the world by making people aware of the cause of so many of the world's biggest and most intransigent problems. He died in 1974, two months after his book won the Pulitzer Prize, but his work has been developed into the field of Terror Management Theory to provide experimental evidence to support his theories.

And this uncomfortable truth about human nature brings us back round to birds. Responding to the subconscious awareness of this buried fear of mortality by seeking out "symbolic immortality" is a fundamental driver of human nature, Becker wrote. One way we do that, Dickinson writes, "is to project power and importance onto some idealized other, often a celestial god . . . something larger that will save us." Birds are also potent symbols that buffer the anxiety born of our awareness of our own mortality, which is likely a big reason behind the passion for bird-watching: "Not quite celestial, they have the unusual capacity to take to the sky with a beauty, mystery and charisma that renders them elusive, godlike and apart from us," she writes. "These characteristics make them ideal symbolic

'transference objects' on which to project a striving for immortality."
Research shows that dreams and fantasies of flight are linked to feelings of transcendence of death.

Perhaps these things are what drove the writer and birder Phoebe
Snetsinger. In her book *Birding on Borrowed Time*, she describes
noticing her first bird in 1965, at the age of thirty-four, a Blackbur-
nian warbler, which is black-and-white with an orange splash on its
throat. As she looked through her binoculars she was astonished. "I
thought 'My God, that is absolutely beautiful.'" The bird sparked
something deep inside her. "Here was something that was happening all of my life and I'd never paid attention to it. It was as if a window opened up." In 1981, just before she turned fifty, she was "given
a death sentence," she wrote, diagnosed with a fatal melanoma with
six months or so to live. She began spending her family inheritance
chasing birds across the globe. (Coincidentally, she scored number
8,024, a northern pygmy owl, with Denver Holt, after she contacted
him about where to see one. "We spent two days together, and found
a nest and watched a pair copulate and sing and enter and exit the
nest hole," Holt told me.) In spite of the fact that she was attacked in
New Guinea in 1986, and suffered numerous other injuries, she was
the first person to break 8,000 and eventually reached about 8,400
before she was killed in a van crash in 1999 while on a birding ex-
pedition in Madagascar, searching for a little bird called Appert's
Greenbul.

I thought about how birds lift us out of our earthly concerns when I
went to see the owls with Denver Holt, the owl field biologist. His
office and home are on an old farm next to the federal Ninepipe
Wildlife Refuge, near the tiny speck of the town of Charlo, Montana, on the Flathead Indian Reservation.

Holt is athletic and compact, in good physical shape from running and hiking up and down hills chasing birds for the last few de-

cades. He has a round, ruddy face, creased by a life outdoors and frequent laughing, and a youthful enthusiasm about his work and life in general.

We drove one day to capture some long-eared owls for research, and Holt told me, in his rapid, clipped New England accent, how he ended up as an owl researcher. "Birds of prey changed my life," he says. One of two kids, he grew up in South Boston. In his rebellious teen years, he fell in with a partying crowd who used to skip school to do drugs and drink. Holt also had a penchant for stealing books, and his house was full of hot volumes. "From everywhere, anywhere. Libraries, stores, you name it. Mostly books on nature, the American West, and Indians. I stole lots and lots of books. I even ripped the 'Word Power' out of every single *Reader's Digest* that I saw, to try and teach myself a better vocabulary." A professor persuaded him to return the books, but he still has the Word Power pages.

Along with "Denver the troublemaker," as many people who knew him thought of him, there was an inquisitive kid eager to learn, with a penchant for nature. His father was raised on an Oklahoma farm, which Denver visited often, spending his time wandering the fields and forests, watching foxes, deer, and birds. And in Massachusetts he liked to do his partying in the woods. He and his buddies even built a log fort there big enough to park a Volkswagen, so they had a place to hang. One day a woman named Nancy Claflin called Holt at home and told him, ever so politely, that his party-fort was a fine and interesting structure but had been built on property belonging to the Massachusetts Audubon Society and he would need to take it down. "And if you ever need a summer job," she added helpfully, "it looks like you've got a lot of good energy. Maybe we have something for you."

The party fort came down, and early the next summer Holt went to Claflin's upscale home, rang the doorbell, and reminded her about her job offer. Even though she had visitors, he says, she came out to talk with him and told him to come back with two letters of recommendation. "And I did, from a truant officer and a cop. They said, essentially, this kid is running around crazy, but he has a good heart," Holt recalls. So Claflin hired him to do landscaping.

Working in Claflin's yard one day he saw a large bird swoop across the sky. He turned his head up to watch it and noticed the sunlight glowing through translucent red tail feathers.

"What are you looking at?" Claflin asked.

"A red-tailed hawk," Holt said.

"How do you know what it is?"

"Well, it's got a red tail," he said. "That's not hard. We used to go to the country with my dad when I was a kid and I was interested in hawks."

Holt thought nothing of it, but when he reported for work a few days later, Claflin said, "Let's go for a ride." He got in her car, and there on the seat were a brand-new pair of 7x35 Bushnell binoculars and a guide to the birds of North America by Chandler Robbins. "She said, 'These are for you,'" and they drove to a site to watch peregrine falcons that had been reintroduced by the Massachusetts Audubon Society. Holt looked through the binoculars and saw the bird world's most impressive hunter cut through the sky. His interest in birds began to take over his life, and gradually he moved away from the ground-level activities of stealing books, running numbers, and partying hard.

A devout bird-watcher herself, Claflin took Holt under her wing and tutored him on the subject of birds. She arranged a job for him at Mount Auburn in Boston, a sprawling, historic cemetery famed for the bird-watchers who gather there, especially during the spring, when thousands of birds migrate through on their way north, and he began to befriend and to learn from the serious birders there.

After graduating from high school, Holt crisscrossed the country in a pickup truck, always with his binoculars on the seat beside him, "keeping an eye on birds." In the fall, he started school at Massachusetts Bay Community College, where he played baseball, as both pitcher and third baseman. "I had guys on the baseball team watching hawks flying over during practice," he said.

Holt left school in Massachusetts and journeyed to Montana in 1976 to finish college, and it was during his junior year at the University of Montana that owls came into his life. Someone told him about an owl nest in a tree on the edge of town, so Holt and a classmate, Billy Norton, went to check it out. "We looked up, and there was a pygmy owl and a saw whet owl in the same tree, and I thought 'Wow, this is so *cool*!' We skipped most of our classes and took shifts to observe them, and we jotted notes down, every day through the spring of that year. Just me and Billy, two young kids who didn't know what the heck we were doing." One of his professors said, "This is the kind of energy and spirit we should have," and he helped guide the two burgeoning ornithologists.

"While we were sitting there looking at those owls every day, I thought 'These guys are so *cool*, just so *cool*," he recalls, excitement still in his voice forty years later. "Look at them, look at their behavior, their appearance. Every time I looked at them I thought, What a cool-looking animal, this is one of the coolest-looking animals out there.' Everyone was studying grizzly bears and wolves and eagles, and no one was doing owls. What an opportunity, an opportunity to learn." He'd found his totem. He and Billy Norton published their first paper—"The Simultaneous Nesting of a Pygmy and Saw Whet Owl"—in a journal called *The Murrelet*. "That was it, I was hooked."

Holt's grades weren't good enough to get him into graduate school in Missoula, so he worked wildlife jobs part time and researched and wrote papers on the side. With twenty papers written, he applied to graduate school at the University of Montana again in 1987—and was again rejected. His grades were just too low. "I still had a thing for owls and decided that if the university wouldn't take me, I'd start my own research institute, the Owl Research Institute. So I got three buddies to be the board of directors. Now it's known worldwide, and we probably have eighty or ninety papers under our belt." John Fitzpatrick of the Cornell Lab calls Holt "one of the premier owl researchers in the world." In 1998 Holt bought the farmhouse and outbuildings that now house the Owl Research Institute. It includes a small, rustic writing cabin with a stunning picture window view of a bird-filled wildlife refuge and the snow-mottled Mission Mountains. It's called the Nancy Claflin cabin.

Holt approaches science the old-fashioned way, with painstaking, long-term studies that require trapping lots of birds, banding them, and then catching them again the next year, which provides critical data that are not available from the satellite collars or DNA techniques often used these days. The day he and I trapped long-eared owls we drove to some grassy hills, hiked a couple of miles, and at the bottom of a ravine, amid thickets of gnarled chokecherry trees, set up mist nets made of very lightweight filament to trap the flying birds. After the net was set, Holt hiked to the top of the ravine, which flushed two of the owls. They immediately flew down the ravine, zooming low to the ground, weaving amid the trees for cover—until wham, they hit the soft nets. As we scrambled to reach them, we could see two birds calmly entangled, their wide owl eyes unblinking as we slowly unwrapped them and affixed small bands to their legs.

For the last twenty-five summers Holt has been living in a crude cabin near Barrow, Alaska, trapping snowies, studying the swarms of chubby lemmings they swoop down on and eat, examining the feather-lined depression that is their ground nest, counting their downy, googly-eyed young. He has pulled apart tens of thousands of owl pellets, gray or black papery wads that owls regurgitate after a meal, which contain undigested bits and bones of voles, lemmings, or whatever else was for dinner.

And when snowies come to Montana he studies them there. On the day I went out with him, we walked on icy dirt roads below the gray lattice of microwave towers where the birds sometimes perch. Holt picked up an owl pellet, which he said likely contained the remains of five voles. He pulled the wad apart, took out his binoculars, and looked through them upside down at the pellet. He immediately spotted a tiny vole bone. "Hmm, a fibula and a tibia," he mumbled. "And here's a humerus," he said as he pulled out a leg bone the size of a paper match. "Ought to be a skull in here somewhere." And sure enough, he pulled out a tiny vole skull, whose molars held the key to learning what type of vole the owls were eating.

Forty years into his career, Holt's enthusiasm for owls burns as brightly as it did in 1976. Nothing compares to walking across the wind-whipped tundra and watching a snowy owl grab a lemming, he says. At sixty-one he now feels the arduous field days a bit more, especially after hiking ten or fifteen miles across the tundra, but he says he can still scramble up a tree better than most students. "Some of my professors and friends and mentors sat in their office their entire career," he said. "I said, 'Oh man, I don't want to do that.'"

As we stand watching the snowy owls on the rooftops in the cold of winter, Holt is wondering aloud about these magnificent creatures. "All this stuff is going through my head about adaptations," he says. Snowy owls are believed to have split off from great horned owls some four million years ago and staked out new territory in the Arctic. To survive in their new home in the extreme far north, they had to evolve, actually change their physiology. Their especially thick coat of feathers, for example, keeps them warm in subzero temperatures and their unusual white color camouflages them. Some traits, though, are still a mystery. "Why have yellow eyes? Why!? What purpose does that serve?

"And how can they stand Arctic habitat, hunting in twenty-four hours of darkness, with snow on the ground, and then switch over in summer to hunt in twenty-four hours of light?" he asks. "What are the adaptations for that?

"I am still fascinated by all of this stuff, just fascinated," Holt says as he packs up his spotting scope, and one of the birds turns to watch as we head toward the car. Denver Holt, a troubled teenager pulled out of a downward spiral into a career as one of the world's leading owl researchers, is nowhere near retired. Owls and the unanswered questions about them got hold of him, and they aren't letting go. "They say the more you know, the less you know. It's true. The older I get, the more questions I have."

It's clear that birds have the ability to lift us, to heal our very beings, to give lives meaning and purpose. The inner city of Washington, D.C., holds a remarkable example of how we can apply the transformational power of birds.

# CHAPTER 18

Birds as Social Workers

*I believe I can fly, I believe I can touch the sky.*

—R. KELLY

Rodney V. Stotts would be dead, or in jail for the rest of his life, had he not met Mr. Hoots and Harriet. Stotts was living the criminal life in Southeast Washington, D.C., in the 1990s, making good money selling crack cocaine and "love boat," which is marijuana laced with the hallucinogenic drug PCP. It's a perilous way to make a living, and dozens of Stotts's fellow dealers were gunned down or stabbed, or had overdosed. "The funeral home made a lot of money off of us," he says, laughing darkly. A banner year for the funeral director was 1992, when thirty-three of his compatriots were killed on Southeast D.C.'s mean streets.

Everyone calls Stotts "Bird Man." I talked to him at a barn he was rebuilding to house his birds. He is tall and whip thin, and as he loped over to his van, where we could sit and talk, he kept a watchful eye on his two restless pit bulls. Stotts, who is Muslim and wears a *taqiyah,* or prayer cap, sat in the driver's seat and pulled on a cigarette as he told me in his raspy smoker's voice that his childhood in Valley Green was classic Anacostia, "drugs and guns, bodies outside every day, people getting shot." His mother worked at a dry cleaner's, and about his father he says he remembers only two things. "He bought me a Popsicle once, and he used to beat me. That's it. Never played cards, never shot baskets, never did anything. He died when I was sixteen."

The same year that thirty-three of Stotts's fellow drug dealers died, Stotts did nine months in prison. Upon release he needed paycheck stubs to prove gainful employment in order to rent an apartment. A new group doing environmental cleanup needed volunteers, he'd heard, and though there was no pay, they offered a stipend.

So on a spring morning in 1992 Stotts showed up with eight others at Anacostia's Lower Beaverdam Creek. He was in an alien world, wearing boots that reached his hips and pulling old tires, broken furniture, and other trash out of what was called "the filthiest creek in America."

Several things pulled him out of his descent into violence and drugs, including this and other cleanups the Earth Conservation Corps worked on, Stotts says, but at the heart of his change were the birds he started training for the group, in particular the raptors he fell in love with. He now raises several different kinds. "I fly the birds *every* day," he says emphatically, taking a drag on a cigarette, savoring the smoke for a few seconds. "I *hate* bad weather because I can't fly. I love it when there's a breeze, then I fly birds all day. If I had forty birds I'd fly my birds all day long, not do anything but fly my birds. Take my birds out. Get peace."


To the Romans, Calidarius was a legendary bird the color of snow that was able to take on someone's sickness, fly away, and expel it, leaving the person, and himself, well. The idea that animals can heal is more than myth, though. Pets have long been known to help people feel safe and comforted; therapy dogs, therapy cats, even therapy hedgehogs are an integral part of the health care scene these days; and dogs and horses are used to rehabilitate prisoners and troubled youth.

Again and again as I researched this book I heard stories about how birds had transformed people in profound ways. In addition to Denver Holt, I've met Vietnam vets who found that the flying of golden eagles helped relieve them of their war trauma; and I read about the famous Birdman of Alcatraz, Robert Stroud, who raised canaries in his prison cell and journeyed from a life of violent crime to one as a respected ornithologist who, from his prison cell, published scientific papers on bird diseases. In the tough, crime-ridden neighborhood of Compton, California, dozens of black and Latino former gang members developed an obsession with Birmingham racing pigeons, raising and racing them competitively, and they became the subject of the documentary *Birdmen*. "It's like a form of therapy for all of these people," Milena Pastreich, the filmmaker who directed it, told *Time* magazine. And it's not just anecdotal; studies done by the Research Center for Human/Animal Interaction at the University of Missouri College of Veterinary Medicine and the Human/Animal Interaction program at the National Institutes of Health show significant gains from animal-assisted therapy.

But by a long shot, the most impressive story I have ever heard about the ability of birds to mend trauma-ravaged lives is the one I encountered on the banks of the Anacostia River, in the neighborhood of the same name in Washington, D.C., where Rodney Stotts came of age, when these streets were infamous for crack-fueled violence and an astronomical murder rate. An unusual and beautiful bird culture has found purchase in this unlikely place and has deeply transformed many lives. Now Stotts and his falconry mentor, Robert Nixon, are trying to bottle this bird lightning to bring its power to bear in the lives of thousands of Washington's disconnected youth, including hundreds of juvenile prisoners who came off the same mean streets in one of the country's most troubled neighborhoods. One day, they hope, it can be exported to other cities.

When I first heard about Bob Nixon's raptor therapy I tried to contact him, without success. It turns out that he is wildly busy—he is not only involved with the Earth Conservation Corps and a group he and Stotts founded called Wings Over America, he is also a documentary filmmaker and was just finishing a film called *Mission Blue*, about the octogenarian oceanographer Sylvia Earle and her campaign to draw attention to the decline of the world's oceans. When we finally connected he was heading off to Africa to shoot another documentary, *The End Game*, about elephant poaching. Still, he graciously agreed to meet me for a few days of show-and-tell, and said we could also meet the D.C. police chief and some other police brass and detectives.

"Police?" I asked. "What do they have to do with birds?" He didn't answer, but I found out soon enough. Nixon also offered to find a nearby hotel room for me, and when he couldn't, he and his wife, Sarah, invited me to stay at their home in Georgetown, a rambling eighteenth-century mansion, one of the oldest in the city.

In his early sixties, Nixon still has boyish, almost cherubic looks, graying strawberry-blond hair, and a sturdy, compact frame. Like Sekercioglu, the biologist from Turkey, who is trying to head off bird extinctions, he is pained by a natural world under assault, having been committed to wildlife issues since he was a boy. He grew up a child of privilege in a Philadelphia suburb; his father, Robert, was a Chrysler executive, and his mother, Agnes, creator and head writer for *One Life to Live*, *All My Children*, and other soap operas. Young Bob, who hung out with animals in the fields around his home, wanted to study wildlife biology. "Then I discovered someone put math in biology, and I flunked it," he said. When he was seventeen, Nixon's father took him to England to meet Phillip Glasier, one of the world's top falconers and the proprietor of the Falconry Centre, a tourist attraction and bird breeding facility. He stayed for a year to learn the sport.

As a young falconer, Nixon flew birds for films, then segued into filmmaking himself to indulge his passion for protecting endangered species by capturing them on film and making people aware of their plight. One of his first films, the Academy Award–nominated *Amazon Diary,* explored the mystical relationship between the Kayapo Indians of Brazil and the harpy eagle, and the tribe's efforts to protect the birds' home in the forest.

The relationship between Nixon and a young crew of inner-city conservation volunteers was sparked in November 1991 in Malibu, California, in a small home, a "shoebox on stilts" so near the Pacific Ocean that the surf crashed over it during big storms. Over a mug of coffee in his kitchen on a sunny California morning, Nixon read an article in *The New York Times* about volunteer efforts to clean up Lower Beaverdam Creek, a tributary of the Anacostia River. A billion dollars had gone into restoring the Potomac, complained Norris McDonald, the head of the African American Environmentalist Association, who was quoted in the piece, and "virtually none of that money went to the Anacostia River."

The color of chocolate milk and just shy of nine miles long, the Anacostia—sometimes called the "black river" because it flows through the African American neighborhood of Anacostia—meanders from southeast Maryland to its confluence with the Potomac River, also known as the "white river," which weaves through primarily Caucasian neighborhoods. Washington, D.C., sits between the two rivers, while the neighborhood of Anacostia sits across the river to the east of the capital.

For decades the Anacostia has been written off as little more than a dumping ground, its banks choked with old tires, abandoned cars, occasional bodies, and raw sewage that washes in during heavy rains. And Lower Beaverdam Creek was the worst of the worst.

Nixon was intrigued by this tale of environmental injustice because of a promise he'd made to a friend. In 1982, he and a line of camera crew and porters macheted their way through a tangle of cloud-shrouded jungle and tall grass to a collection of rustic tin-roofed cabins high on a saddle between two volcanoes in Congo's Virunga National Park; they were there to profile a then little-known primatologist named Dian Fossey for ABC's *American Sportsman*. Fossey, a groundbreaking researcher and a fierce advocate, was maintaining a lonely, armed vigil over the planet's last two hundred or so mountain gorillas to keep them from being wiped out by poachers. She was wildly eccentric, with a reputation among locals as a witch who could cast spells. As Nixon and his crew finally reached her cabin, she appeared on the porch brandishing a shotgun, wanting to know what the hell they wanted; she had forgotten they were coming.

During a week of pouring rain in the cloud forest, over interviews and drinks at night, the two became fast friends, and Fossey told Nixon about the book she was writing about her life among the mountain gorillas. Nixon said her fight to protect the animals would be a great subject for a feature film that could greatly help her cause. She shook her head. "I promise you no one will be interested in my story."

Before he left, she offered to give him the film rights to her book if he promised to spend a year of his life in an active conservation project. " 'You come here from your fancy New York pad, spend ten days here, and you're gone,' " Nixon recounted her telling him. " 'I want you to do a hands-on project, not a film project.' "

"So I agreed," he told me.

When he returned home, Nixon shopped around the idea of a feature film on a crusading primatologist and discovered Dian was right: No one in Hollywood was interested. That changed in 1985, when Fossey was murdered. She was found in her bed, hacked to death by a machete, likely by the poachers she'd battled, though the killers were never caught. Her story was made into the 1988 film with the same title as her biography, *Gorillas in the Mist*, with Nixon as a co-producer.

A few years later, still looking for a way to fulfill his promise to Fossey, Nixon read the *New York Times* article on Lower Beaverdam Creek and decided he'd honor it by organizing a cleanup. When he arrived in 1992 to recruit his first nine volunteers, he thought it would be a parachute drop. "I figured I'd start the project, spend a year, point out the problem, and the cavalry would arrive, and I'd go back to making feature films," he explains. "But the cavalry never arrived."

Nixon toggles back and forth between two very different worlds. Since the late 1970s, he has made his living making documentaries in order to fund his environmental campaigns, and in 1988 he shot a documentary for ABC called *America the Beautiful*, about America's most scenic places. He asked then President George H. W. Bush to host it. To everyone's surprise—except his, Nixon says—Bush agreed. In 1990, as Nixon filmed Bush fishing for tarpon on the Florida Bay, he asked the president about a public-private partnership the White House had created called the Earth Conservation Corps, billed as "a million kids planting a billion trees." The program had never gotten off the ground. Might he take over the unfunded corps, Nixon asked, to raise money and recruit volunteers for cleaning up the Anacostia? Talk to that guy over there, the president said, pointing to Marlin Fitzwater, the White House press secretary, and the White House agreed.

Nixon sought his first nine Earth Conservation Corps recruits in Anacostia. "I asked 'Where is the scariest place in Washington?' And everybody said 'Valley Green.'" Southeast D.C. was the murder capital of the country in those days, making up the bulk of some five hundred homicides citywide a year (as opposed to about one hundred now), and the housing project with the halcyon name of Valley Green was ground zero. Many in the senior class of the high school here had been murdered before graduation. With help from locals, Nixon called a recruitment meeting in the Valley Green community center, and dozens of would-be volunteers showed up. "I showed them pictures of the creek, I showed them the *Times* article, and

I said 'This river belongs to you, who wants to volunteer to clean it up?' Nine young men and women said yes." Rodney Stotts, still involved in the world of the streets, was one of the original nine. Volunteers received a twenty-dollar-a-day stipend and a five-hundred-dollar scholarship when they finished. "We attacked that creek," said Nixon.

This was not to be a Disney-style troubled-street-kids-become-volunteers-for-the-planet-and-all-is-now-well-with-the-world story. Their efforts continue, to this day, to be a traumatic journey through a grim and endemic gauntlet of violence. Nixon was taken aback. "It dawned on me that there was a third-world country six blocks from my house," he said. Or, as Stotts put it to me, "This is like Beirut," and then, pointing across the river, where the Capitol dome gleamed, he said, "That's like Beverly Hills." Nixon required that everyone put their weapons in the trunk before they started out on their environmental restoration projects. "Sorry," one mumbled, sheepishly pulling out a Smith & Wesson Model 57, "had to go out and use the phone last night and forgot I had it."

The first serious violent episode took place in 1993, not long after the corps's founding. After the crew mucked out Lower Beaverdam, the U.S. Department of the Interior asked them to come to Texas to stack sandbag barriers to help protect a brown pelican rookery from erosion in Big Boggy National Wildlife Refuge. The group sublet apartments in nearby Houston. On the second night, a young man who had leased out one of the apartments returned and raped and killed nineteen-year-old corps crewmember Monique Johnson. Later they discovered she had been pregnant by Gerald "Tink" Hulett, another ECC volunteer. Everyone was devastated.

Still, the volunteer work continued, even as more members of the crew were violently killed—stabbed, shot, beaten. In the year after the group was formed, three of the first crew of nine were killed by street violence in their neighborhood.

By 1994 Nixon had sold his shoebox house on the beach in Malibu in order to help keep the corps going and was living full time in

D.C. One day he met an elderly man named Julius Lowery walking along the Anacostia who told him bald eagles used to soar along the river, but in 1947 they had abandoned their nest. Pollution, especially DDT, which thins eggshells and thwarts successful hatching, had wiped out the nation's symbol in the shadow of the Capitol. The conversation gave Nixon an idea.

With the Anacostia River growing cleaner, and more fish calling it home, Nixon and the crew decided the natural next step in the restoration process was to bring back the top-of-the-food-chain predator, the bald eagle. Some biologists laughed and told him that it would never work on such a polluted waterway, but Nixon persisted.

Bald eagles are among the biggest of birds; their wingspan reaches more than seven feet and they weigh ten to fourteen pounds, as much as a bowling ball. They hunt fish and small animals and scavenge on carcasses as well. They almost disappeared across the Lower 48, dwindling to just 450 nesting pairs by the 1970s, and they were listed as an endangered species. With the help of a ban on DDT, captive breeding, and other restorative measures, they've come soaring back, and now there are nearly ten thousand pairs.

First up for ECC's reintroduction project was scoping out a likely eagle tree. In a donated boat, Nixon and the crew cruised up the muddy Anacostia one afternoon until they spied the silhouette of a perfectly suited old-growth oak tree, taller than all the others, on the grounds of the 444-acre National Arboretum. They landed near it and hiked into the arboretum offices—which must have seemed odd to the staff, Nixon told me, laughing, because in the mid-1990s nobody came in from the river, especially young black people, clothes wet and covered with mud. They explained what they were after and were ushered in to meet the acting director, who, to their surprise, readily agreed to allow the crew to place their first "hack box," a wooden cage where the birds could grow up, in the big riverside oak.

Bald eaglets are often born in batches of three, and in a fierce competition for food, the first- and second-born will often peck the youngest to death or push it out of the nest. Biologists sometimes remove these third birds, called "terminal eagles," before they are killed and use them to create new populations. Nixon applied to the U.S. Fish and Wildlife Service to acquire some—a far from certain prospect, he said, as his office at the time was in the backseat of his Volkswagen Jetta. But he knew people at the Fish and Wildlife Service through his film projects, and his connections helped earn the ECC four fledgling bald eagles. The first football-size chicks were taken out of their nests in Wisconsin and brought to D.C. in the spring of 1994.

The crew built their hack box—a four-by-eight-foot plywood box with wooden dowels on top and in front—and placed a nest inside. With a giant crane used to repair power lines, the Potomac Electric Power Company lifted the box into the branches of the tall tree. The gangly eaglets were placed in the nest inside, and the babysitting and worrying about the birds began.

To keep the young birds from being fooled into thinking the corps members were their parents, the crew stayed out of sight and kept careful watch with binoculars and spotting scopes. "Those kids never took their eyes off those birds," says Nixon. "It was an impressive effort." Three times a day, every day for six weeks, a bucket full of fish parts was hoisted by rope and pulley to the top of the tree above the nest and then shaken, until a smelly, delicious rain of food fell onto the eaglets.

On a sunny day in mid-July, when the birds were old enough to fledge, the crew yanked a rope and the front of the box sprang open. The birds looked around, tentatively flapped for a while, and then finally flew away from their tree, their untested wings growing more powerful with each stroke. The corps members stood mesmerized as the now massive birds—*their* birds—soared over the Anacostia. Suddenly and spontaneously they broke into R. Kelly's then popular song, "I Believe I Can Fly." Over four years they released four birds each year, two of them wearing satellite tracking devices, and all of them banded.

The three young African American conservationists who were killed during the first year of ECC's effort were not its only casualties; since 1992, twenty-three volunteers have met a violent end on the streets of Anacostia—killed by friends or strangers. A young eagle's life is also perilous; the mortality rate for juvenile bald eagles is about 70 percent. The uncertain fate the birds and conservationists shared created a deep feeling of kinship. "That's why, when our eagles were ready to be released into the wild, we named them after our fallen members," Stotts explained to me, perhaps a spontaneous, modern manifestation of the widespread ancient belief that human souls live on in birds after death. The first was named Monique. The second was Tink, after Monique's unborn child's father, stabbed to death over a ten-dollar debt. Benny Jones was named after a young man beaten to death with a length of pipe for not moving off a park bench. Diamond Teague's namesake was shot and killed as he stood on his front porch. "What was shocking to me was for kids to be murdered and their names weren't in the paper," says Nixon. "Not even an obituary. And that's one of the reasons we started naming the birds after them, to honor them, because nobody else did." The ECC also created a media arts team that made a documentary called *Endangered Species*, about the lives of their fallen members.

Four years later a pair of bald eagles came back to nest along the river. It was the first time in fifty years that the eagles had nested there, and the crew was ecstatic. The corps members were convinced they were Monique and Tink come home. Two years later a second pair built a nest on the grounds of the Washington, D.C., police academy.

I first heard about this eagle project from Cornell's Janis Dickinson, who piqued my curiosity when I read about it in her paper on terror, birds, and immortality. "Each eaglet is an immortality symbol named after a young corps member murdered in the difficult neighborhood," she wrote. "The eagles are a symbolic perpetuation of these young lives and, as a consequence, solidify the ideology that holds the corps together." The birds, in the language of Terror Management Theory, are symbols of immortality: powerful, soul-deep diversion from the deep fear and anxiety of daily life and death in Anacostia. What might perpetuate the violence and drugs and poverty in such inner-city environments, in fact, is generation after generation growing up amid such fear and trauma.

Our fear of death weakens our self-esteem, and a basic tenet of TMT holds that humans seek to bolster their self-esteem as a buffer against existential anxiety. The "cultures of honor" that form in response to coming of age in an especially harsh environment, such as urban gangs, stave off fear of death by providing a place for members to gain honor, albeit honor achieved through violence or other sordid activities. Many traditional societies, Becker, the anthropologist, writes, understood this dynamic and nourished the need for heroic ventures in warfare or hunting. The Earth Conservation Corps, Dickinson told me, is a positive way to create a similar culture of respect that restores and protects the world's most charismatic "immortality symbols" and, in the process, supports the heroic culture of the corps.

The pinnacle of this achievement may have been at a ceremony on the White House lawn in July 1999 when twenty or so ECC members were guests of President Bill Clinton to celebrate the bald eagle's remarkable recovery from near extinction and its removal from the endangered species list. Levar Sims of the ECC introduced the president in a moving talk about working with the bird. These eagles "made me think about life and what it means to be endangered," he said, and he explained that he had named one of the eagles LB after a slain friend, Leroy Brown. But Sims said the program had saved his own life. "I was given back to my community to teach young people what I know about the Anacostia," he said.

While raising four batches of eagles was a remarkable success for the ECC, as the last four flew off in 1998, Nixon told the crew that it was the project's final year. "Even though that was the plan all along, as soon as the last eagles fricking flew off, the corps members were devastated," he told me. "They had raised them, named them, and loved them, and now they were gone." He needed to bring birds permanently into the program, he realized, for the sake of the crew and many others whose broken lives might be mended with the help of feathered friends. Nixon had come into his own through the bird window, "and it was obvious that it could be a way for them," he said.

As we chatted about this over lunch in Anacostia, I asked Nixon what he thought it was that gave the big birds the power to change lives. He shook his head slowly for a few moments, at a loss to come up with a way to express it. "Birds are just magical in all they can do," he said finally. "They are free, they fly, and they fit so seamlessly into their world. They are kind of like perfect beings. And bringing them to the corps would allow these kids to be part of something bigger than themselves."

So in 1999, a New York falconer and breeder Nixon had met at the Falconry Centre donated Mr. Hoots and Harriet to the Earth Conservation Corps. Mr. Hoots is a Eurasian eagle owl, a big and dramatic bird that is several shades of brown with intense orange eyes and a regal presence, while Harriet, a mix of rich, mahogany brown feathers, with a penetrating stare, is a sleek Harris's hawk. Harris's hawks are known for hunting in packs, like wolves. This would be the beginning of the Earth Conservation Corps's use of trained birds of prey to help both the birds and the people who raise them lift themselves up.

Falconry, the sport of kings, immediately began to emerge as a powerful force among these young inner-city men and women. Rodney Stotts began training with Nixon to learn to fly birds and eventually became a falconer. As he started taking the ECC birds out to fly, he told me, he felt something inside him stir. "Whenever something tough was going on in my personal life, I got a bird and walked," he said. "I found that it really gave me some clarity, I had time to think. I loved birds especially because they are wild in the beginning, and then you catch them, train them, tame them, and you have them flying free to you. Birds can take off at any moment, but this bird loves you enough to come to you, trust you. That's the part that was for me."

Another graduate of the program is Robert West, so good with raptors that Nixon reverently refers to him as "the bird whisperer." He served in the ECC in his teens and now, at twenty-five, is a master falconer. He's moved out of the projects and works at the American Eagle Foundation, in Pigeon Forge, Tennessee, helping care for some eighty birds, many of them sick and injured. "I found my love for birds of prey," he told President Barack and First Lady Michelle Obama at a ceremony at the Lincoln Memorial during the 2008 inauguration, looking very much at home with Challenger, a bald eagle, on his arm, "and I have been pursuing it ever since."

When D.C. police chief Cathy Lanier watched West at the inauguration with Challenger on his arm, she was thunderstruck. "It's hard to explain," she told me, "but seeing a bald eagle up there on that stage with the president of the United States, and the first African American president at that, was just beautiful. Then, when the handler turned the bird toward President Obama, the eagle spread his wings. It was incredibly moving, the best part of the whole inauguration for me." Lanier and her department later worked with Nixon and Stotts and Wings Over America. And in what has to be a first, the entire D.C. police force has adopted bald eagles as their animal totem. When I met West, I asked him why he thought these birds were so appealing to inner-city young men and women. "They're free," he said simply.

There were birds, big raptors, always around us in D.C. for the three days I was there. One morning I came down for breakfast at the Nixons' and heard the cry of a bald eagle as we were drinking coffee. It was Challenger, the bald eagle from the inauguration ceremony, in a dog kennel next to the kitchen table. And I felt that this was as it should be, to always be in the presence of such extreme natural beauty and power. Challenger is owned by the American Eagle Foundation, run by Al Cecere, and he travels around the country, flying at football games and banquets to raise funds for the foundation. The corps also takes in and rehabilitates wounded raptors. Once they become healthy they're released, though birds that can't survive in the wild—such as Harriet, who is missing a talon—are kept for show-and-tells in schools around Anacostia.

I went along on one road show before a lively gathering of fourth and fifth graders at Our Lady of Perpetual Help, a Catholic girls' grade school, where the students dress in prim plaid uniforms. The ECC's Darrell Wallace had brought the cinnamon-colored Harriet out of a dog carrier to the gasps of thirty or so young ladies who were suddenly all eyes. As he removed a small dead mouse from his pocket, the squeals began. When Harriet rapidly gulped down the whole mouse in a few seconds, swallowing the long thin thread of the tail last, the squeals grew higher in pitch. As the excitement died down, Wallace opened the floor to questions. "Is it true," one young lady asked breathlessly, "that eagles carry off small children?"

I asked LaShauntya Moore Reynolds, another corps graduate turned staffer who helps display these birds in schools, why she thought these big birds were so appealing. "It's a sense of freedom," she reiterated. "These birds are doing exactly what you need to do in your life to overcome things, to soar, so the birds symbolize that."

One of the nests the reintroduced eagles built is atop a tree on the grounds of the police academy. "There was a real sense of pride that it was not only in the nation's capital, but on the academy grounds that a pair of bald eagles nested," says John Mein, a community outreach officer with the Criminal Intelligence Branch of the D.C. police force, who climbed a precarious oak tree to mount a webcam trained on the nest. "A lot of the officers, as they have gone through training, doing sit-ups and push-ups, they see these eagles flying overhead. So there's a feeling among the force that they're 'their' birds," he says. The eagle's nest camera was first switched on the day after the Boston Marathon bombings, and as the eagles appeared on a screen before an emergency meeting of the police command staff, everyone applauded. "It was the highlight of the week for these lieutenants and commanders and the whole police force," says Mein. Chief Lanier named the first two eaglets in the nest Liberty and Justice.

Osprey have also returned to the river, and there are now twenty-five nesting pairs along its shores. In 2013, Chief Lanier helped biologists trap and radio-tag the birds, one of which she named Rodney, after Stotts. The other was named Ron Harper, after the Pittsburgh steelworker who is father of the Washington Nationals' standout baseball player Bryce Harper, a local hero.

The D.C. police command center now has two large computer screens; one displays the real-time whereabouts of some fifteen hundred juvenile felony offenders who, because of the condition of their probation, wear GPS trackers, and the other displays the whereabouts of Ron Harper and Rodney as they soar over the east coasts of North and South America on their migration.

As Mein and I stood on a busy street looking up at the huge pile of branches that is an eagle's nest, he told me these birds and others like them have a unique power to reach young offenders. "There's an immediate connection with these birds," he says. "I've seen it. It has a calming effect, and the kids are totally in tune with what Rodney's saying." Mein is working closely with Nixon and Stotts. "With the right structure and resources, Rodney could train a whole crew of falconers from D.C.," Mein says.

Lanier has also used the big birds to lift spirits. When Nixon told her that Challenger was in town, she asked a favor of Cecere, the founder of the American Eagle Foundation, who cares for the bird. Hundreds of khaki-clad police cadets were scouring a park for an eight-year-old girl, who police believed had been brutally murdered and buried. Could Challenger make an appearance to lift the morale of the troops taking part in the grim search? Cecere's daughter Julia arrived at the site of the search, took the big bird out of the back of her van, and undid his leather hood to display his regal visage. The response was immediate and dramatic. Dour faces turned into smiles and then grins, cameras came out, and "oohs" and "aahs" were heard. "It was emotionally draining for everybody, because for days we searched every inch of that park for that little girl," Lanier said. "But birds have a way of lightening your spirit, and brightening the mood. And it worked. It moved people. It did wonders." Grievously, the girl's body was never found, but a small measure of peace was brought to the community that day.

Nixon drove me out to a forested spot near Laurel, Maryland, at the headwaters of the Patuxent River, where he and others are now betting big on the power of raptors to transform. In 2009 he started Wings Over America, which is comprised of just himself and Rodney Stotts, and they have grand plans for this 880-acre former dairy farm that was also the site of the D.C. city government's youth detention facility, which was closed in 2004. The district donated the land and buildings to Wings Over America, and the group plans to turn it into a national center for birds of prey run by at-risk youth who are serving time.

The verdant acreage where Wings will be based is studded with large old trees and is bookended by two facilities—New Beginnings Youth Development Center, a prison for juveniles who are violent offenders, and the Capitol Guardian Youth Challenge at the other end of the property, where young people at risk voluntarily enter a military-style boot camp which can lead to a GED. By healing, feeding, and teaching birds to fly again, Nixon and Stotts hope that people from both groups will be inspired and enabled to leave behind turbulent, dead-end lives. While they are preparing the birds for a return to the wild, the birds are preparing the young men, hopefully, for a meaningful and successful return to society.

Now forty-five, Stotts is the lone staff person at Wings. He, his cousin, and some of the detainees are painstakingly cleaning out and rehabbing the dairy barn, the greenhouse, and other ramshackle abandoned buildings that once made up the farm, and they are constructing new aeries for the couple of dozen raptors they are planning to house here. They are also building flight cages—eighty-foot-long, twenty-foot-wide, fourteen-foot-tall wire-fence structures. Many groups treat injured birds by placing pins in a broken wing or leg, but they don't have space for long-term flight rehabilitation, which is vital for a successful return to the wild. Now, once the birds are patched up, Wings will bring them here and teach them to fly again.

Because of his own drug-dealing, gun-toting past, Rodney Stotts has a lot of street cred with young black offenders, who come to trust him. "You find the ones that are terrified at the beginning, and by the end of the day they have an owl on their arm," he told me. "You see that light turn on."

After Stotts and Nixon showed me their bird facilities, I chatted with Rodney. I asked what his friends from the old days might think of his life with birds. "Friends? I don't call anyone walking the earth my friend," he shot back. "I don't believe in that word. The only person who can hurt you is someone who you love, right? That's why I don't have friends, because they can't hurt me if I don't love them, right? I trust the birds more than I trust any person. If a bird puts its talons in you, you understand that because that's what they do. They are wild animals. Humans are supposed to be civilized, but why are we doing things worse than wild animals? People who you show the utmost respect to, they turn around and stab you in the back."

In 2010, after studying with a falconer, Stotts passed a test to receive his state and federal falconer's licenses. When the state inspector came to look at his mews, or bird pens, he told Rodney that in thirty years on the job he was the first black falconer he had inspected. With his falconer's license Rodney was now ready for his own wild bird. He set out a *bal-chatri* trap, an ancient device first used in East India, nearly as old as the sport of falconry. The trap is essentially a cage covered with wire loops; inside the cage Stotts placed a pigeon. "Any bird that is flying over and sees it, if he is hungry enough, comes down for a meal," Stotts explained to me. "Hopefully it catches his foot and then it's too heavy to fly off. If it's a juvenile, there's your bird."

The first bird snared in the loops of Bird Man's *bal-chatri* trap was a kestrel. "The smallest raptor in North America, a female, and she was beautiful," he said appreciatively. "Because she was a female and it was a female that was murdered first, I named her Monique"; just like the bald eagles they released, this bird became another namesake of the young pregnant friend who was killed in Texas. His second bird, a red-tailed hawk he named Tink, was also named after someone who died; all his birds have been, save one. "That one's name is Moochie," he told me. "That's my little sister. She has lupus and stuff like that, so instead of giving her flowers when she was gone, I wanted to give something to her while she was here. So I named one of my birds after her."

The open, parklike campus in Maryland is the perfect place for Rodney to fly his birds and maintain his equilibrium. "My uncle passed three weeks ago," he said. "I went out and flew my birds. That's all I do. Make my *salat*—that's a prayer—and go fly my birds. Whatever it is going on slowly melts away."

What Rodney Stotts, Bob Nixon, Chief Lanier (who recently left the department to head security for the NFL), and others have found is a vital but still a faint glimmer of this connection, a very rich source of emotional energy. The dynamic relationship between people and nature is essential and often miraculously therapeutic. Birds have become important to the D.C. police and others living amid quotidian urban terror because, as research shows, they provide us with robust relief, and even, for a time, transcendence, from our most fundamental fears and anxieties. So many of the world's problems, especially our disconnection from nature, is caused by these deep-rooted anxieties. As Denver Holt showed us in the previous chapter, and as we see here in the story of an inner-city falconer and his partners, reconnecting people with nature has the ability to heal both us and the world. How can we take this knowledge and build on it even further?

On the morning before I left D.C., Bob Nixon and I sat in his high-ceilinged library, crammed with books, arrows from his trip to the Amazon, and other assorted memorabilia and art, and he again brought up the subject of bird magic. "It's not that mysterious if you've ever fed a bald eagle on your fist and had it fly to you," he said. A large measure of the transformation that has happened to a handful of people along the Anacostia, he said, "is being able to care for something as outrageously wild and beautiful as a bald eagle, our national bird. These are kids who, in their own words, are 'America's nightmare,' and to be given the federally permitted custodial responsibility for the nation's bird in the shadow of the Capitol is so powerful.

"It's a tale of two cities, as extreme as it ever has been," he continued. "They are pulled over all the time by the police, they are shot by their friends and colleagues. It's such a tragic waste of talent. These birds gave them the ability to dream and to think they can do something positive, it's as simple as that. A lot of people here say these birds saved their life."

If we can learn how to move beyond the subconscious terror we all carry and the emotional numbing we take on to shield ourselves, if we can tap into the extraordinary power of birds and bottle this lightning, if we learn from our relationship with birds to fully understand our nervous system and the full range that we are capable of feeling and sensing in the world, we will find something inexhaustible and profound, even life-changing. "Wilderness holds answers to questions man has not yet learned to ask," said the writer Nancy Newhall. Learning to ask those questions about how we might do things differently is the heart of the task before us. And a small, dedicated group of ornithologists is leading the search.

# CHAPTER 19

⌒⌒⌒◦⌒⌒⌒

# Expanding Our Senses

No voices now speak to man from stone, plants and animals, nor does he speak to them believing they can hear.
His contact with nature has gone and with it has gone the
profound emotional energy that this symbolic connection
supplied.

—CARL JUNG

The Yaghan people, a tribe of hunters and gatherers that once
roamed the rain-shrouded Cape Horn region of Chile, tell a story of
a brother who was deeply attracted to his sister. Despite the taboo,
the boy tried every which way to sleep with her. She refused him,
though secretly she harbored similar feelings. One day the boy told
his sister of a place with delicious giant red berries, the kind she
loved to pick, called *chauras,* and she took her basket and hiked off

into the forest to find them. The brother had gone on ahead and was hiding in some dense brush as she passed, and when she neared he stood up and embraced her. Fueled by pent-up desire, they fell to the ground and made frenzied love. The moment they finished, legend has it, they were turned into male and female *lana*, the Yaghan word for the Magellanic woodpecker, the big, showy insect hunter that lives in the forests of southern South America. That's why the male *lana* to this day carries the red crest, the Yaghan people believe, a reminder of those berries the brother used to entice his sister.

Tales of birds turning into humans and humans becoming birds are very common among the world's indigenous peoples. The story of the *lana* has several meanings for the Yaghan people. It teaches the young about a taboo relationship that could reduce genetic fitness. It's also an origin story about an ancestral time when humans and birds were one, which speaks to the close relationship between birds and the Yaghan people.

That story was told by Cristina Calderon of Puerto Williams, Chile, a tiny outpost at the bottom of the world near Cape Horn. Calderon is the last full-blooded Yaghan and the last native speaker of the Yaghan language. The wizened eighty-nine-year-old, known as *abuela*, or grandmother, makes a living weaving and selling traditional baskets and is paid by tourists who disembark from the mammoth cruise ships that dock near Puerto Williams to pose for a photograph or to utter a few sentences in her all but moribund language.

The Yaghan people of Tierra del Fuego, the southernmost people in the world, were once nomads who lived in sealskin tents and paddled in sealskin canoes between the cliffs of the narrow, rain-soaked fjords and among islands, hunting seals, sea lions, and otters and gathering shellfish. They were among the last native tribes of South America to encounter European invaders, and when they did they paid dearly for it. In the late eighteenth century, preconquest, there were several thousand Yaghan. As the Europeans brought disease, forced their removal, and even slaughtered them, the Yaghan numbers spiraled down until, in 2005, Ms. Calderon became the lone survivor.

When she dies, all native speakers of the Yaghan language will be gone, and along with them much of the unique Yaghan perspective on the world—ecological, cultural, and biological. And birds play a significant role in this perspective. So central is everything about birds to the native way of life here, as it is among many other indigenous tribes around the world, that a primary method for documenting this language and culture is through their stories of birds.

Throughout the world and across history, a great many native peoples and birds have been deeply connected. In countless myths they transform into people and vice versa, and they have long been conflated with angels and with other mysterious beings, and have been thought to have special powers. Nearly four thousand years ago, in a shallow sandstone canyon in the sere desert of southwest Texas, for example, artists left dozens of detailed multicolored depictions of a cloud of soaring birds guiding shamanic figures who were half bird and half human. The dominant theme of these paintings, according to the archaeologist Solveig Turpin, who has studied them, is "the bird-like flight of the soul." Bird motifs throughout prehistoric art, in fact, represent the ethereal magic of shamanic journeys, of birds as psychopomps, spiritual creatures who guide souls from the earth to the afterlife. Most native cultures have lived among birds, worn their feathers, eaten them, used them for divination and rituals, ascribed magical powers to them, and mimicked their calls with their songs and language. They figure prominently in almost all native creation stories. Indigenous people don't only live near birds, they emotionally relate to birds as fellow travelers, sometimes even as a kind of extended family. So if we seek to understand how to build a better relationship with birds and nature, we should look to the world's indigenous cultures for some answers.

That's precisely what the field of ethno-ornithology does. This tiny discipline within the field of anthropology studies "the complex of interrelationships between birds, humans, and all other living and non-living things, whether in terrestrial or extra-terrestrial spheres or in body or in spirit," writes the anthropologist Eugene Hunn in the definitive book on the subject, *Ethno-Ornithology: Birds, Indigenous Peoples, Culture and Society.*

The field was born in the 1970s when traditional ornithologists working with tribes realized that these cultures related to birds in ways vastly different from our own, and that in studying the biology of birds they were getting only a small, distorted picture of a rich, ancient, and holistic relationship that had evolved over thousands of years. So the ornithologists brought together other disciplines—linguistics, anthropology, and cognitive science, as well as traditional natural science methods—to arrive at an almost unimaginably different understanding of the interconnectedness of birds, people, and the natural world.

Understanding the many different ways people live with birds offers essential wisdom for our own culture. "The most important question we face today is a question of ethics," says Ricardo Rozzi, a professor of biology at the University of North Texas, who, with his wife, Francisca Massardo, a fellow conservation biologist and anthropologist, studies the bird tales of the Yaghan along with another indigenous Chilean culture, the Mapuche. "The Yaghan see birds as living beings with an intentionality and interest in life. They are not just passive objects, machines for producing meat. Lamentably, however, that is how birds are seen by many. By culturally encountering the birds, though, we can have a better life and the birds can have a better life. To the Yaghan, birds are companions and teachers, very much."

Our disconnection from the rich relationship human societies once had with the natural world was called "the Great Forgetting" in *The Story of B,* by Daniel Quinn, the 1996 novel about a priest who discovers that the natural world has a spiritual aspect. Along with biodiversity, biocultural diversity—the planet's vast, millennia-old relationships between people and nature—is disappearing as dominant cultures and rampant resource development decimate and extinguish many of the last natural environments and the people living in them. Nearly half of the six thousand languages that once thrived on the planet will be gone by 2050—disappearing at four times the extinction rate of birds. Along with them goes the deep knowledge of plants and animals that they express, which is the product of thousands of years of evolution.

The anthropologist Wade Davis calls the constellation of native cultures who have accumulated wisdom and learned to thrive in their landscape for generations the "ethnosphere," or "the sum total of all thoughts and dreams, myths, ideas, inspirations, intuitions, brought into being by the human imagination since the dawn of consciousness. It's a symbol of all that we are, and all that we can be, as an astonishingly inquisitive species." These myriad perspectives manifest in language, which is not merely vocabulary, he writes, but "a flash of the human spirit, the vehicle by which the soul of each particular culture comes into the material world. Every language is an old-growth forest of the mind, a watershed of thought, an eco-system of spiritual possibilities."

Preserving a language, then, is preserving the spirit of a people. Because birds are so charismatic and so intricately woven through the lives of indigenous people, ethno-ornithologists believe the best way to perpetuate the essence of these cultures is to catalog their stories of birds.

As this knowledge disappears, the work is increasingly vital. There are perhaps a hundred ethno-ornithologists in the world, a tiny but global effort. Oxford University is home to the Ethno-ornithology World Archive, an online clearinghouse for video interviews, films, depictions of rituals about birds, studies, stories, testimonies about how things have changed, photos, and other bits and pieces of bird culture. Preserving ancient knowledge is why Rozzi helped found Omora Park in southern Chile, dedicated to preserving both the biodiversity and the cultural diversity of the Yaghan and Mapuche people. And it's why he coauthored a multiethnic bird guide that combines scientific understanding of birds such as the Magellanic woodpecker, austral parakeet, and steamer duck with the cultural perspectives of the indigenous people.

Birds have also been enlisted to help preserve the worldview of two all-but-extinct cultures on a tiny spit of land called the Andaman Islands, located in the Indian Ocean in the vastness between the coast of India and Myanmar. Five groups of pygmy peoples called the Andamanese may be the most isolated people on the planet, and their population has dwindled to just a few hundred souls. One of these groups, the hunting-and-gathering Sentinelese, with a couple of dozen members, know nothing of the modern world, and another, the Great Andamanese, are just a few dozen people who speak their own unique language. The authors of *Birds of the Great Andamanese* gathered their bird stories as a window into those cultures' dwindling ways of life.

The seminal work in the field of ethno-ornithology is *Birds of My Kalam Country,* written by Ian Saem Majnep, a member of the Kalam tribe from Papua New Guinea, who was born before Europeans first came to the remote Kaironk Valley. This small country is one of the most linguistically, biologically, and culturally diverse places on the planet, with four and a half million people speaking some eight hundred languages. In a study of his people, Saem detailed the traditional perspective of 137 species of bird and six species of bat (which are considered birds by the Kalam). His book provides in-depth descriptions of birds and bird behavior, as well as their role in hunting, magic, mythology, and ceremonies.

Many of the world's tribes have their own elaborate and distinct taxonomy. The Cheyenne of the Great Plains, for example, divided birds into three groups: holy, great, and ordinary. Each group occupied a different level of the cosmos. The highest, including ravens, crows, and woodpeckers, soared in what the Cheyenne called the Blue Sky Space above everything else. Magpies were a special bird because they spent time among people and overheard conversations that they reported back to the high gods and to a spirit named Sweet Medicine, a Cheyenne prophet. The great birds—many eagles, hawks, buzzards, and short-eared owls—lived in the Near Sky Space, and the ordinary birds—robins, turkeys, waterfowl, kingfishers, and meadowlarks—flitted about the Atmosphere on the lowest level.

Cheyenne healers treated diseases and injuries with bird parts. Feathers from a kingfisher were used to heal wounds, notably bullet wounds, because of the way water "heals" after the aerodynamic bird dives to catch a swimming fish.

Butterflies and dragonflies, called "whirlwinds," were part of the Cheyenne bird group, but among the owls only the short-eared owl was considered a bird, because it was active during the day. All other owls were called *mista,* or "spooks of the night." The bald eagle was not considered a true eagle because it eats fish, unlike the others that consume only meat. It was also regarded as a thief because it steals fish from osprey. So it made sense to the Cheyenne that the thieving eagle was a symbol of the federal government, because they felt that the United States had stolen from, and mistreated, native people.

Rozzi's multiethnic bird guide describes the similar traditional views of Yaghan and Mapuche bird culture, but it also includes basic scientific understandings. Take the green-backed fire crown, a hummingbird, Omora Park's namesake. The small bird's color, size, and other physical attributes are described in English in the guide. Then alongside that is the bird's name in Yaghan, *omora,* which means "little spirit" and reflects the belief that the bird leads humans to safety during times of trouble and maintains harmony between society and nature. Then in Portuguese, the guide tells us, the hummer is called *beija-flor,* or "kiss the flower," while in Spanish it is *picaflor,* or "pierce the flower," which refers to what the bird must do in order to drink its nectar, while in Mapuche the bird's name is *pinda,* which means "humming." Bird sounds and songs are also included on a disc that accompanies the book.

Bird legends, too, are recounted in the guide. The tribe members were suffering from severe drought, the Yaghan hummingbird story goes, and a mean fox guarded the only lake that still held water. Despite the people's thirst, the fox stubbornly refused to share the water. Omora asked the fox to please, please give the humans some water, but the fox laughed and sneered and said, "Go away, tiny bird." The fox turned his back, though, and the wily hummingbird stabbed him with a harpoon and then ripped down the fence so the people could reach the water. This David and Goliath story, says Rozzi, explains the Yaghan belief that all life is connected and that cooperation is essential for survival. "There is a duty to share," says Rozzi. "Private property is called into question always."

Sharing is a common theme among indigenous bird stories, because it's a matter of survival. A clan of the Bosavi people, the Kaluli, who live in the forest foothills of a collapsed volcano, Mount Bosavi, in the interior of Papua New Guinea, have a tale of the *muni* bird with a similar lesson. A brother and sister go fishing for crayfish, and the sister catches some but doesn't give any to her younger brother. This is a violation of basic social etiquette. The brother takes a crayfish shell and puts it over his nose, which turns purple, and then he turns into a purple-nosed fruit dove, or *muni,* and when he opens his mouth to speak, his voice turns into a bird's voice. This becomes a form of crying with poetry in it, a profound expression of loss and abandonment. The bird's song becomes a metaphor, then, for this deeply felt emotion that is fundamental to the Kaluli. "The girl's action snaps the thread of social bond," says Steven Feld, an anthropologist at the University of New Mexico who has studied the Kaluli and their birds for twenty-five years. "Their deepest fears are those of loneliness, a breakdown in reciprocity, and not sharing raises feelings of loss, vulnerability, abandonment, and death."

Bosavi stories that are essential to social cohesion are performed as highly complex, emotional songs that are the imagined journey of a bird through the valleys and forests of central Indonesia. "Say someone in my family passed," Feld says. "You would sing a song to me about a certain creek or tree, and then name a certain quality of light and the sound of a particular kind of water and the sounds and places where I've gardened, where I've hunted, and where members of my family trapped or hunted. These songs aren't sung from the point of view of someone walking this route, but from the point of view of a bird, so they are actually maps of all the bird flight ways." Each song begins with the name of a bird.

The vital songs and stories of bird flight, in a culture without a written language, are also a very powerful way of keeping biography and history, Feld explains. "It's also the equivalent of a land deed, because it tells which families traditionally used those landscapes. 'My ancestors always went there, and we have these songs that show it.' When you plot the seven thousand points in Bosavi stories with a GPS on a map, they are all separate, and so these stories are a poetic cartography of the forest." Now, as Exxon prepares to build a gas pipeline through the thick carpet of forest in Bosavi territory, they are negotiating easements with members of the tribe based on their bird poem maps.

The Bosavi live with 125 species of birds, and the tribe regularly interacts with 85 or 90 of them. A large tract of forest that surrounds the village is off-limits to hunting, so birds know they are safe there, flocking, nesting, and roosting alongside the tribe. These birds provide a deep and continual source of knowledge about the goings-on of the world, as well as entertainment and companionship. So tuned in are the Bosavi to birds that when they hear birdsong, "they immediately know the time of day, the season, the layer of forest canopy the bird is in, what fruits are in season, and they would know if it rained two hours ago, and other things. All of this eco-knowledge is instantly available," Feld explains. Birds are also considered ancestors, carrying the spirit of departed relatives. "You're not just listening to a bird in the tree, you are listening to Uncle Charlie," Feld says. The people still shoot birds and eat them, but not the ones they believe to be their relatives.

There's a tale about two fish swimming side by side in the sea. An older fish happens by and says, "Morning, boys, how's the water?"

The two young fish look at each other and say, "Water? What's water?" Consciousness is, essentially, our awareness of the world, the water we swim in all of the time without realizing how differently others see the world. These very different relationships to birds don't just tell us something about birds, they tell us of radically different kinds of consciousness, of a multitude of alternative ways of perceiving and of being. Yet we are so steeped in our own ways of swimming in the water that it's hard to even imagine such things.

In his 1985 essay "The Perceptual Implications of Gaia," the philosopher David Abram describes an alternative view of perception, very different from the way our culture defines it. "Traditionally, perception has been taken to be a strictly one-way process whereby value-free data from the surrounding environment is collected and organized by the human organism," Abram writes. "The external world is tacitly assumed to be a collection of purely objective, random things entirely lacking in value or meaning until organized by the ineffable human mind."

That's wrong, he argues. Relying on the work of two experts in perception, Maurice Merleau-Ponty and James J. Gibson, Abram believes that communication between an individual and the planet's vast assortment of life is "an interactive phenomenon." Psychologists have erred by studying perception in the lab, he writes, and if they studied the phenomenon in nature, "they would come to understand the senses not as passive mechanisms receiving valueless data, but as active exploratory organs attuned to dynamic meanings already there in the environment." The world, in other words, is fantastically rich and alive with meaning and feeling independent of us, but for most of us, the real failure is our inability to sense it.

This is an extremely important idea. My own reading and reporting on perception tells me that as Westerners we perceive too much with our eyes and ears and fail to move our awareness into our bodies, to literally feel the world around us. "If one is successful in this, then one may abruptly experience oneself in an entirely new manner—not as an immaterial intelligence inhabiting an alien, mechanical body, but as a magic, self-sensing form, a body that is itself awake and aware from its toes to its fingers to its tongue to its ears, a thoughtful and self-reflective animate presence," Abram writes. "Birds, trees, and even rivers and stones begin to stand forth as living, communicative presences." Nature, to us in the modern world, is something we think of as out there beyond our sealed, air-conditioned houses, our office buildings and cities and automobiles. Without our technologies, many other cultures have adapted to nature very differently—by immersing themselves in it, becoming part of it.

I have experienced a small sense of this very different way of being in the world. When I spend long periods in nature on back-packing trips through wilderness, I feel far less separate from the world, more at home in it. My perception is enhanced: Sunlight seems to sparkle, and the scarlet color on a hepatic tanager or the blue of a lazuli bunting look richer and more dramatic, while the perfume of flowers smells more intense.

There are ways for us to take this deep and robust sensory connection with nature further. A friend and coauthor of mine, Les Fehmi, a psychologist and veteran Zen practitioner, told me a story of attending a *sesshin*, a week of intensive meditation, from dawn until evening each day, a practice that sensitizes the nervous system and enhances experience. As he sat in a restaurant one early evening during that week, in front of a plate glass window, he watched as white gulls floated over a Southern California bay. "All of a sudden I was a gull," he told me. "I could feel my feathers and wingspan, how they cut through the air, and I could hear and feel myself calling. Everything was magnified. It was wonderful. I had adopted totally, for a few minutes, that mode of being."

This ability to reach out beyond the body and intimately sense the world is "a heightened form of mindfulness," Felice Wyndham tells me. Wyndham is an anthropologist and ethnobiologist who works with the Ethno-ornithology World Archive in Oxford and studies how different indigenous peoples relate to landscapes. "It's quite common, you see it in most hunter-gatherer groups," she says.

"It's an extremely developed skill base of cognitive agility, of being able to put yourself into a viewpoint and perspective of many creatures or objects—rocks, water, clouds. We, as humans, have a remarkable sensitivity, imagination, and ability to be cognitively agile. If we are open to it and train ourselves to learn how to drop all of the distractions to our sensory capacity, we're able to do so much more biologically than we use in contemporary industrial society. It's probably some kind of very early cognitive development, a prerequisite for sociality, the ability to see yourself mirrored in your peers. You learn it very early in the attachment process with your mother, and then you detach from your primary relationship, and continue with other people. If you are in a highly diverse and sensuous natural environment, you are also going to be doing that with all of the organisms, the plants, the water, and the birds—especially the birds—which is exciting, because they fly and this gives you a completely different perspective."

This is what Ricardo Rozzi of Omora Park is getting at when he talks about the Yaghan of southern Chile and their relationship to birds. One of his native collaborators is the Mapuche poet Lorenzo Aillapan, who, Rozzi says, has a rapport with the birds of southern Chile because of these different sensory capabilities. "When I walk with him through the forest and he plays his instrument, the birds respond, they answer him," says Rozzi. "It's a dialogic relationship. It's similar to the Franciscan tradition of communicating with the birds." Saint Francis of Assisi was known for walking among birds, who listened to him, stayed close to him, and showed no fear.

These kinds of ideas are why Abram wrote that "The ecological crisis may be the result of a recent and collective perceptual disorder in our species, a unique form of myopia that we are now forced to correct."


When we look at how the Yaghan and hundreds of other cultures relate to the natural world, we realize that our model of relating to birds is only one possibility, just one star in a constellation of many different ways of being in the world. Regarding birds as family, forming kinship with them, singing their songs, wearing their feathers to be closer to the spirit world, believing they are extensions of our thoughts and feelings or messengers through whose eyes we can see the world as they fly, may be just as valid as, or even perhaps more valid than, seeing birds as things to be measured, weighed, counted, named, eaten, and perhaps admired, and nothing more.

Ethno-ornithology tells us, through birds, that there is far more information in the natural world than we know and are able to access. Gathering and protecting this knowledge preserves cultures that have evolved with birds, cultures that have learned to sustain themselves with bounty from the natural world largely without destroying it. Understanding the relationship of native cultures to birds may lead us back to a sustainable world in which their fate—and ours—is no longer in doubt.

# EPILOGUE

*~~~~~*

# The Future of Birds

Wherever there are birds, there is hope.

—MEHMET MURAT ILDAN

"I do not know what I may appear to the world, but to myself I seem to have been only like a boy playing on the seashore," wrote Sir Isaac Newton, "and diverting myself now and then, finding a smoother pebble or a prettier shell than ordinary, whilst the great ocean of truth lay all undiscovered before me."

While we've clearly made huge strides in our understanding of nature since Newton's time, I would argue that the great ocean of truth still lies undiscovered. And it's a perilous time to be so much in the dark. Temperatures and ocean levels, already on the rise, will increase further as the climate changes, and we'll have more, and more intense, hurricanes and other storms, withering drought, punishing floods, and destructive winds. The relentless loss of natural habitats and biodiversity to industrialization and urbanization continues. And our relationship to the natural world is often distant and infrequent.

Fortunately, it's not all bad news. There is a burgeoning realization of the scope of the damage that has been done to the planet, and there is a growing movement of people around the world who are planting trees, tearing up pavement, and bringing nature into cities. And as we have seen time and again, if given a chance, nature mends and restores itself. If we connect and repair fragments of forest, for example, their hospitality to birds and other wildlife is enhanced and they will move back in. Wolves, wiped out of the West by the 1950s, were returned to Yellowstone National Park in 1995 and since have repopulated much of their old range. And many species of birds, from peregrine falcons to bald eagles, have been brought back from the edge of extinction.

The more difficult part of the dilemma we face, though, concerns the things we don't know about how the world works. "Just to know that Jaguar shamans still journey beyond the Milky Way, or the myths of the Inuit elders still resonate with meaning, or that in the Himalaya, the Buddhists still pursue the breath of the Dharma," anthropologist Wade Davis said in a TED talk, "is to remember the central revelation of anthropology, and that is the idea that the world in which we live does not exist in some absolute sense but is just one model of reality."

Some scientists also acknowledge that our basic comprehension of nature is woefully incomplete. Within our traditional scientific model there is a long list of unknowns. We know very little, for example, about information transfer among animal groups. Yet a decline in communication among birds in a flock and a breakdown in the integrity of the flock "metamind" could be responsible for some extincions, both in the past and in the future. We don't know what birdsong and bird calls are about, or what they tell us about the world. And despite years of investigation, we don't fully understand how birds migrate.

Then there are concepts that aren't part of mainstream scientific thought but that many experts believe should be, because they are essential to understanding what is really going on in the world. The notion of "human exceptionalism," for example, is a fundamental assumption that is being reexamined, and the incredible intelligence of birds is one of the chief reasons to question it. "The belief that only humans are capable of experiencing anything consciously is preposterous," writes neuroscientist Christof Koch in *Scientific American*. "A much more reasonable assumption is that until proved otherwise, many, if not all, multi-cellular organisms experience pain and pleasure and can see and hear the sights and sounds of life. . . . And the more complex the system, the larger the repertoire of conscious states it can experience." As we have seen, birds, with their fantastically evolved and detailed nervous systems, and their ability to talk, sing, perform complex thinking, mate for life, and use tools, are among the planet's leading candidates for conscious creatures. Koch writes that the notion that every living thing is conscious, a philosophy called panpsychism, is "the single most elegant and parsimonious explanation for the universe I find myself in." If it's true that we are not alone as conscious creatures, that in itself should compel us to rethink our place in the world, as part of it rather than as its masters, to do with it as we wish.

Another fundamental challenge to our current scientific model is the very real possibility that quantum principles—which, among other things, hold that things that seem unconnected really are connected in a hidden, or unknown, way, and that phenomena that seem random really aren't—are manifest not only in the atomic world but in the larger world around us. If this is so, it would provide the scientific underpinnings to the theory that birds can navigate by seeing magnetic lines or that a human mind can sense or feel the flight of a bird. The world may well be a much stranger and radically different place than we can even imagine. "In recent years, our knowledge of such things has made huge strides—and not only in connection with birds," writes Jim Al-Khalili and Johnjoe McFadden in their book

*Life on the Edge: The Coming of Age of Quantum Biology*. These professors (of theoretical physics and molecular genetics, respectively) argue that we need to accommodate quantum principles into our model of how the world works. "There is no doubt that much of what is or was wonderful and unique about robins, clownfish, bacteria that survive beneath the Antarctic ice, dinosaurs that roamed the Jurassic forests, monarch butterflies, fruit flies, plants and microbes derives from the fact that, like us, they are rooted in the quantum world."

It's intriguing that these new ideas in science are similar to what some indigenous people have believed all along—that all of nature is conscious and deserving of respect, for example. That's why it's important as well to turn to what Wade Davis calls the ethnosphere, that vast repository of information and knowledge that has risen out of thousands of years of humans adapting to nature. Based on what other cultures tell us, it's time to realize that the answers may not lie only in what we do out in the world, in terms of reclaiming and restoring nature, but also in how we perceive the world. What role does fear play in our perception? We really need to know, for example, what "heightened mindfulness," as anthropologist Felice Wyndham calls it, is: how it works and how it might be learned to enhance our sense of the world. Understanding the many different ways of being in the world can help us understand how we can enrich and enhance our own perception and experience of nature.

And all of this leads to the biggest and most important question of all: Do we live in a giant complex and physical machine, made up only of what we can see and measure with the tools we have? Or do we live in a sacred world infused with some kind of spirit or intelligence—as myriad cultures, from Native American to Buddhist, believe and have believed over thousands of years?

It's time to let birds be our guides on a journey back to the natural world, toward a Great Remembering, to counter what the writer

Daniel Quinn called "the Great Forgetting." We are creatures of nature, who have evolved over thousands of years in its midst, and today we are suffering greatly because we have separated from it. We need to follow our feathered friends back to the forests, prairies, coastlines, and mountains, and we also need to make better homes for them and other species in the cities and suburbs, so that they might help us restore that lost part of ourselves, a part that the Bosavi, Yaghan, Yanomami, and many other cultures seem to have never lost.

The first step, then, in reclaiming that relationship with nature is proximity. "Half of our first-year students in biology can't name five British birds, and twenty percent of them can't name one," Andy Gosler of the Ethno-ornithology World Archive at the University of Oxford told me. "That's a common story across the world. When you say, 'Did you know that this species has declined seventy percent in the UK in twenty years?' it means nothing to them. They say, 'Why should I care? I didn't know that bird existed until you told me about it.' But when you take people bird-banding, and put a bird in their hand, it is a life-changing experience for them."

"When I take out the most cynical individuals, and put a peanut in their hand, and a jay hops into it—jays are absurdly tame—there is a connection that is felt by that individual," says Cornell's John Fitzpatrick of his research on scrub jays, which, like white-fronted bee-eaters, form families. "And when I tell that person, 'That bird is eight years old, and he is with his second wife, and he's had four kids and one of those kids is breeding next door, and this bird is going to disappear unless we keep pieces like this managed as habitat,' that's a life-changing experience. But getting them out there is the real challenge."

I have learned from birds that we need new relationships among birds and people that can heal both, like the relationship between the Earth Conservation Corps in inner-city Washington, D.C., and the bald eagles they restored to the Anacostia River. A modern, dysfunctional American city—the nation's capital, no less—creates a birds-of-prey culture to bring back bald eagles and teach falconry and bird rehab to heal its residents. "Can working with birds really make a difference with some of the hardest cases?" I asked former Washington, D.C., police chief Cathy Lanier, who dealt with some of the nation's toughest crime. "Oh, yes," she said. "The kids in trouble are in trouble because something was missing in their life. It put them on the wrong path. Not a hundred percent of the time, perhaps. But for a large majority of them, the things that are missing can be brought back by engaging them in the lives of these birds."

The wonder of birds first dawned for me, and changed my life, on a warm May day in 1980, when I trudged to the top of a treeless grassy hill overlooking the valley sprawl of Boise, Idaho. I was there to interview a man named Morley Nelson, an internationally known Idaho falconer, a fierce conservationist who had trained eagles for Walt Disney films. He was championing protection for the Snake River Birds of Prey National Conservation Area, a rocky river canyon in the Owyhee Desert of southern Idaho with the densest raptor population in the world, about which I was writing my first magazine story.

Then in his mid-sixties, Nelson, a decorated World War II combat veteran with the Tenth Mountain Division, had a sad, stoop-shouldered look about him. He got out of his truck, greeted me, and lifted the rear window. There in the back of a dumpy pickup truck sat one of the bird world's most stunning raptors, a peregrine falcon—at the time an endangered species. It had a slate-blue cowl and a dun-and-white belly and chest. It was a bird with serious gravitas, and this was the first time I had seen one up close. I was entranced. Nelson put on a thick leather glove that covered his forearm and untied the bird from its perch. He moved his arm toward the falcon, which stepped onto it with its powerful yellow talons. The bird wore a chestnut-colored leather hood over its head, which Nelson removed, and the regal bird stared at him with its fierce yellow eyes. Nelson walked the bird to the very top of a hill, removed equipment attached to a leather lace on the bird's leg, and lifted his arm. With a great clatter of its powerful wings, the peregrine took to the sky, corkscrewing up into the silken clouds. In a few minutes it was out of sight.

The peregrine falcon is a powerful predator, and on its dives it can reach speeds of 200 miles per hour or more, slamming into its prey—a duck or a pigeon sauntering along at a top speed perhaps of 50 miles per hour—like a missile. We didn't see it kill anything that day, but we did watch, riveted, as it dove, soared, and wheeled through the sky. Nelson's whole demeanor changed as the bird flew; he seemed to find a current in the bird's soaring that energized him. "Look at that!" he said, a jack-o'-lantern grin on his upturned face as he watched the avian acrobatics. "God damn, I wish I could fly like that!" After an hour or so, Nelson whirled a small piece of meat on a cord in a circle around his head, and in a few minutes the bird spied the bait and returned from the sky to alight on Nelson's arm. Nelson fed the bird more meat from a leather pouch on his belt, covered its head again with the hood, and stroked the velvety brown feathers on its broad chest. "All right! All right! Good job!" he said to the bird in an excited sotto voce, and it seemed to me as if part of him had left the earth with the bird.

That was my first time up close with such a magnificent wild bird, and I, too, felt I had, for a brief time, soared with the peregrine, perhaps the closest I would ever come to flying. As that experience taught me, birds speak to our deepest selves and raise our vision up into the skies, literally and metaphorically. They help us to imagine, and even to feel, what it is like to be weightless. The wonder of birds can bring wonder into our own lives. Just imagine.

# ACKNOWLEDGMENTS

A robust thank you to the bird people—their enthusiasm and love for birds is infectious. Ken Dial, Bret Tobalske, Harvey Karten, Ondi Crino, Dick Hutto, Iain Couzin, Craig Reynolds, Richard Prum, Clover and Joe Quinn, Cagan Sekercioglu, Darcy Ogada, Bernd Heinrich, Thomas Bugnyar, Jeff Lucas, Mike Webster, Steve Nowicki, Stephen Emlen, Patricia Adair Gowaty, Jessica Meir, Ron Rosenbrand, Julie Jedlicka, Stella Capoccia, Eugene Oda, Robert Nixon, Sarah Guinan Nixon, Robert West, Rodney Stotts, LaShauntya Moore Reynolds, Cathy Lanier, John Mein, Felice Wyndham, Andrew Gosler, John "Fitz" Fitzpatrick, Janis Dickinson, Denver Holt, Bob Martinka, Brian Woodbridge, Mark Reynolds, Andrea Cavagna, Erich Jarvis, Ricardo Rozzi, Thomas Malone, Lee Plenty Wolf, Willy Zelowitz, Lelan Berner, Louis Lefebvre, Christof

Koch, and Morley Nelson were all very generous and giving with their time. And much, much gratitude to my editor Cindy Spiegel, my other editor Annie Chagnot, and my agent, Stuart Bernstein. And DD Dowden, thanks for allowing your wonderful bird drawings to grace my pages. May the bluebirds of happiness mob you all.

# SELECTED BIBLIOGRAPHY

## 1. Birds: The Dinosaurs That Made It

"How a New Theory of Bird Evolution Came About." *Science Daily,*
    March 3, 2009.

## 2. Hummingbirds: The Magic of Flight

Morgan, Michael Hamilton. *Lost History: The Enduring Legacy of*
    *Muslim Scientists, Thinkers, and Artists.* National Geographic, 2007.
Robbins, Jim. "Flying Machines, Amazing at Any Angle." *The New*
    *York Times,* January 3, 2011.
White, Lynn, Jr. "Eilmer of Malmesbury, an Eleventh Century Aviator :
    A Case Study of Technical Innovation, Its Context and Tradition."
    *Technology and Culture* 2, no. 2 (spring 1961).

## 3. Canaries and Black-backed Woodpeckers: Birds as Flying Sentinels

Carson, Rachel. *Silent Spring.* Boston: Houghton Mifflin, 2002.

Robbins, Jim. "Paying Farmers to Welcome Birds." *The New York Times,* April 14, 2014.

Weidensaul, Scott. *Living on the Wind: Across the Hemisphere with Migratory Birds.* North Point Press, 1999.

## 4. A Murmuring of Birds: The Extraordinary Design of the Flock

Fisher, Len. *The Perfect Swarm: The Science of Complexity in Everyday Life.* New York: Basic Books, 2009.

King, Andrew J., and David J. T. Sumpter. "Murmurations." *Current Biology* 22, no. 4 (February 21, 2012).

Yong, Ed. "How the Science of Swarms Can Help Us Fight Cancer and Predict the Future." *Wired,* March 19, 2013.

## 5. The Power of a Feather

Hanson, Thor. *Feathers: The Evolution of a Natural Miracle.* New York: Basic Books, 2011.

Sveinsson, Jon. "Real Eiderdown." eiderdown.com/files/eider_article .pdf.

## 6. From Egg to Table
PART ONE: THE CHICKEN

Adler, Jerry, and Andrew Lawler. "How the Chicken Conquered the World." *Smithsonian,* June 2012.

Lawler, Andrew. *Why Did the Chicken Cross the World? The Epic Saga of the Bird That Powers Civilization.* Atria Books, 2016.

———. "Stalking the Wild Ur Chicken." *Audubon,* November/December 2014.

## 7. From Egg to Table
PART TWO: WILD BIRDS

Dean, W.R.J., et al. "The Fallacy, Fact, and Fate of Guiding Behavior in the Greater Honeyguide." *Conservation Biology,* March 1990.

Prosper, Datus C. *Pheasants of the Mind: A Hunter's Search for a Mythic Bird.* Wilderness Adventures Press, 1994.

## 8. The Miracle of Guano

Cushman, Gregory. *Guano and the Opening of the Pacific World: A Global Ecological History*. University Press of Kansas, 2013.

Hollett, David. *More Precious than Gold: The Story of the Peruvian Guano Trade*. Fairleigh Dickinson University Press, 2008.

Romero, Simon. "Peru Guards Its Guano as Demand Soars." *The New York Times*, May 30, 2008.

## 9. Nature's Cleanup Crew

McGrath, Susan. "The Vanishing." *Smithsonian*, February 2007.

Santora, Mark. "Vulture Populations Wane, Poisoned by Man." *The New York Times*, August 26, 2015.

Umar, Baba. "Without Vultures, Fate of Parsi 'Sky Burials' Uncertain." Al Jazeera, April 2015.

## 10: Bird Brain, Human Brain

Jabr, Ferris. "Cache Cab: Taxi Drivers' Brains Grow to Navigate London's Streets." *Scientific American*, December 8, 2011.

Specter, Michael. "Rethinking the Brain." *The New Yorker*, July 23, 2001.

## 11. The Surprisingly Astute Minds of Ravens and Crows

Heinrich, Bernd. *Mind of the Raven: Investigations and Adventures with Wolf-Birds*. Harper Perennial, 2007.

Morgana, Aimee. "The N'kisi Project." 2002. www.sheldrake.org/nkisi.

Mortensen, Eric D. "Raven Augury in Tibet, Northwest Yunnan, Inner Asia, and Circumpolar Regions: A Study in Comparative Folklore and Religion." Ph.D. thesis, Harvard University, 2003.

Weir, Alex A. S., et al. "The Shaping of Hooks in New Caledonian Crows." *Science*, August 9, 2002.

## 12. The Secret Language of Birds

Freeberg, Todd, et al. "The Complex Call of the Carolina Chickadee." *American Scientist*, September/October 2012.

Guénon, René. *The Language of the Birds*. World Wisdom Books, 2007.
Stap, Don. *Birdsong*. Scribner, 2005.

## 13. The Bee-eaters: A Modern Family

Angier, Natalie. "Biologists Tell a Tale of Disappearing In-laws." *The New York Times*, April 1992.
Emlen, Stephen T. "An Evolutionary Theory of the Family." *Proceedings of the National Academy of Sciences* 92, no. 18 (August 29, 1995): 8092–99.
Hively, Will. "Family Man." *Discover*, October 1, 1997.

## 14. Extreme Physiologies: Birds, the Ultimate Athletes

Hawkes, L. A., et al. "The Paradox of Extreme High-Altitude Migration in Bar-Headed Geese *Anser indicus*." Proceedings of the Royal Society B, October 31, 2012.

## 15. Nature's Hired Men: Putting Birds to Work

Jedlicka, Julie, et al. "Avian Conservation Practices Strengthen Ecosystem Services in California Vineyards." *PLOS ONE*, November 9, 2011.
"Owls, Kestrels in the Middle East: Flying Mouse-traps Control Pests Without Chemicals." *Science Daily*, April 24, 2009.
Palmer, Theodore S. *A Review of Economic Ornithology in the United States*. U.S. Department of Agriculture, 1899.

## 16. The City Bird: From Sidewalk to Sky

Dunn, Robert R., et al. "Pigeon Paradox: The Dependence of Global Conservation on Urban Nature Conservation." *Biology*, December 2006.
Humphries, Courtney. *Superdove: How the Pigeon Took Manhattan and the World*. HarperCollins/Smithsonian Books, 2008.
Jerolmack, Colin. *The Global Pigeon*. University of Chicago Press, 2013.

## 17. The Transformational Power of Birds

Robbins, Jim. "Getting Wise to the Owl, a Charismatic Sentry in Climate Change." *The New York Times*, May 23, 2011.

Snetsinger, Phoebe. *Birding on Borrowed Time*. American Birding Association, 2003.

## 18. Birds as Social Workers

Barringer, Felicity. "Washington Journal; In Capital, No. 2 River Is a Cause." *The New York Times,* December 1, 1991.

Becker, Ernest. *The Denial of Death*. The Free Press, 1973.

Cohen, Florette, et al., "Finding Everland: Flight Fantasies and the Desire to Transcend Mortality." *Journal of Experimental Social Psychology* 47 (2011): 88–102.

Dickinson, J. L. "The People Paradox: Self-esteem Striving, Immortality Ideologies, and Human Response to Climate Change." *Ecology and Society* 14, no. 1 (2009).

## 19. Expanding Our Senses

Abram, David. *The Spell of the Sensuous*. Pantheon, 1996.

Moore, John H. "The Ornithology of Cheyenne Religionists." *Plains Anthropologist* 31, no. 113 (August 1986).

Rozzi, Ricardo, et al. *Multi-Ethnic Bird Guide of the Sub-Antarctic Forests of South America*. University of North Texas Press, 2010.

Tidemann, Sonia, and Andrew Gosler, eds. *Ethno-ornithology: Birds, Indigenous Peoples, Culture and Society*. Routledge, 2010.

## Epilogue: The Future of Birds

Al-Khalili, Jim, and Johnjoe McFadden. *Life on the Edge: The Coming of Age of Quantum Biology*. Crown, 2014.

Koch, Christof. "Is Consciousness Universal?" *Scientific American,* January 1, 2014.

# INDEX

## About the Author

JIM ROBBINS was born and raised in Niagara Falls, New York, but has lived in Montana since 1977. He has written for *The New York Times* for more than thirty-five years, on a wide range of topics but with a special focus on science and environmental issues. He has also written for *Audubon, Condé Nast Traveler, Smithsonian, Vanity Fair, The Sunday Times, Conservation,* and numerous other magazines. He has covered environmental stories across the United States and in far-flung places around the world, including Mongolia, Mexico, Chile, Peru, the Yanomami Territory of Brazil, Norway, and Sweden.

This is his sixth book. His first, *Last Refuge: The Environmental Showdown in the American West* (1993), was about reconciling the way we as a species live with our knowledge of ecosystems. He is also the author of *A Symphony in the Brain: The Evolution of the New Brainwave Biofeedback* (2000) and co-author of *The Open Focus Brain: Harnessing the Power of Attention to Heal Mind and Body* (2007), about the critical and overlooked role that attention plays in our lives, as well as *Dissolving Pain* (2010), about the role of attention in pain. His interest in the nexus between the human central nervous system and the natural world grew out of these three books.

His fifth book, *The Man Who Planted Trees* (2012), is about the crisis in the world's forests caused by climate change and resource development.

Facebook.com/writerjimrobbins
Twitter: @JimRobbins19